Facing Epistemic Unc

Lay out: Roel van Goor, Amsterdam
Cover design: René Staelenberg, Amsterdam
Cover illustration: Frieda Heyting, Rolde

ISBN   978 90 5629 720 6
e-ISBN 978 90 4851 806 7 (pdf)
e-ISBN 978 90 4851 807 4 (ePub)
NUR   740 / 840

Printed and bound by CPI Group (UK) Ltd, Croydon, CR0 4YY

Facing Epistemic Uncertainty

Characteristics, Possibilities, and Limitations
of a Discursive Contextualist Approach
to Philosophy of Education

ACADEMISCH PROEFSCHRIFT

ter verkrijging van de graad van doctor
aan de Universiteit van Amsterdam,
op gezag van Rector Magnificus,
prof. dr D.C. van den Boom
ten overstaan van een door het College voor Promoties ingestelde
commissie, in het openbaar te verdedigen in de Agnietenkapel der
Universiteit
op donderdag 25 oktober 2012, te 12.00 uur

door

Roel Leonard Christiaan van Goor

geboren te Ospel

Promotiecommissie

Promotor:	Prof. dr. M. S. Merry
Co-promotor:	Prof. dr. em. G. F. Heyting
Overige leden:	Prof. dr. G. J. J. Biesta
	Prof. dr. D. A. V. van der Leij
	Prof. dr. J. Masschelein
	Prof. dr. P. Smeyers
	Prof. dr. J. W. Steutel

Faculteit der Maatschappij- de Gedragswetenschappen

CONTENTS

voor Lize

I
INTRODUCTION: A BRIEF RECONSTRUCTION OF THE DEVELOPMENT OF PHILOSOPHY OF EDUCATION AS AN ACADEMIC DISCIPLINE

1. Introduction

Increasing doubts over what some writers term the 'metanarratives of modernity' (cf. Lyotard, 1984) within Western philosophy have also brought the significance of philosophy to educational thinking into question. These 'major stories' traditionally served to legitimize philosophy of education's - and other societal institutions' - tasks and ways of fulfilling those tasks. "If the Enlightenment idealist and humanist narratives have become bankrupt and must be abandoned...", Peters and Lankshear propose. "wherein can legitimacy reside?" (Peters & Lankshear, 1995, p. 11).

These developments have met various responses amongst philosophers of education. Several authors hold that the radical doubts raised a way of thinking "that puts everything 'up for grabs' (cf. Usher & Edwards, 1994. p. 26). Others respond less dramatically. Whilst, for example, Blake, Smeyers, Smith and Standish (1998), too, think that doubts concerning philosophy of education's traditional justificatory frameworks have led to an 'intellectual paralysis', they do not regard this as the end of the possibilities of educational thinking. They rather approach it as a challenge "to find new resources for thinking again" (Blake et al., 1998, p. 5).

However the developments described above may be interpreted, philosophy of education - as a discipline - is clearly under pressure, which has raised questions about its central tasks and possibilities. These questions are at the core of this dissertation in which I will be examining the recent uncertainties in philosophy of education - in relation to proposals for their resolution. This, for the most part against the background of the reception of recent insights in epistemological thinking.

The practice of questioning its own *'raison d'être'* is nothing new to philosophy of education. The discipline has always - especially since it became detached from an independently developing empirical educational discipline - been engaged in (re)formulating its central tasks and philosophically acceptable manners of fulfilling those tasks. The continuous investigation of philosophers of education into the 'methodo-logy' of their discipline need not surprise us. As Heyting shows, relating "methodological considerations to fundamental epistemological questions" (Heyting, 2001, p. 1) has always been characteristic of philosophy in general. Raising questions about its core business and methods in the light of new epistemological developments appears to be at the very heart of philosophy.

In the remainder of this introduction I will briefly deal with a number of historical approaches to what are considered the central tasks of philosophy of education, and the epistemological developments that can be associated with these approaches, in order to offer a historical background for and theoretical framing of the research questions that are dealt with in the next chapters.

2. The continuous (re)formulation of the tasks and possibilities of an independent philosophy of education

In the early years of its establishment as an independent academic discipline in the nineteen twenties, Dutch academic pedagogy was characterized by a strong philosophical and normative orientation (Mulder, 1989, p. 14; p. 247-48). Philosophy was considered central to educational thinking, formulating the basic principles to serve as point of departure for the construction of educational theory and, from there, educational practice. These basic educational principles were, for the most part, grounded in - either religious, or secular - ideology. What was considered to constitute 'good' education, or upbringing, ultimately depended on personal belief systems, it was thought. Hence, a 'good' pedagogy needed to take belief-systems as its starting point. As stated by the Dutch philosopher of education Waterink: "Particularly in pedagogy, which deals with fostering human beings into ideal adults, towards-an ideal outlook on life, moral standards are paramount to academic endeavor [transl.: RvG]"[1] (1959, p.191-192).

However, in the course of the 20th century a change set in. The 'scientific model' of academic endeavor - derived from the natural sciences - came to be regarded as the model for all academic disciplines; and so also for the behavioral sciences (Slife & Williams, 1995, p. 177). The success of the natural sciences - that had enabled the building of bridges, the treatment of diseases, and the genetic modification of plants - led to a situation in which sound academic endeavor was increasingly regarded as synonymous with natural science. Educational science, following psychology, focused ever more on this 'scientific model', which eventually led to the development of the increasingly independent

---

1 In Dutch: "[J]uist bij de paedagogiek, waar het gaat om het vormen van mensen tot de ideale volwassenheid, tot de ideate levenshouding, daar moet juist deze normatieve situatie in de wetenschappelijke activiteit een eerste plaats hebben".

discipline of empirical educational science (cf. among others Levering, 2003, p. 95; Van IJzendoorn, 1997, p. 78; and 2002, p. 29).

As empirical educational science gained independence, developing its own methods for the provision of scientifically substantiated answers, educational philosophy developed as a more or less autonomously operating discipline with an international orientation. However, the matter of what could be expected of this so-called 'philosophy of education' was under discussion from the very start, not in the least within philosophy of education itself. From the moment it became detached from empirical educational science, the (re) formulation of philosophy of education's central tasks and ways in which the discipline might fulfill these, in a philosophically substantiated manner, appeared to be one of its central themes.

It appeared obvious that questions concerning the values of education, at the very least, belonged exclusively to the domain of philosophy of education. Such questions were evidently beyond the realm of empirical educational science. The second undisputed area of expertise concerned the investigation of matters of a conceptual nature, i.e. of the language of education. Empiricists were engaged with how 'reality' *is*; how that 'reality' is discussed was left to philosophy. The central tasks of the field of philosophy of education, then, seemed clear. How, precisely, these tasks were to be interpreted was, however, not that clear at all.

Philosophers of education who welcomed and promoted an empirically oriented educational science, turned in the main to philosophical approaches such as the 'Wiener Kreis' logical positivism, that sought to arrive at unambiguous philosophical claims by means of rigorous linguistic and logical analyses. Their aim was to settle philosophical issues once and for all through freeing language from its conceptual ambiguities, metaphysical lumber and logical inconsistencies. In their view, philosophy of education was to abstain from metaphysical speculation on the nature and content of education. Because metaphysical speculations could not be rigorously tested through the available and methodologically accepted philosophical apparatus, they were considered meaningless. Philosophers of education, it was believed, should only 'make use of the one infallible philosophical tool: logic, and restrict themselves to the study of the sole philosophically accessible object of research: language. Empirical educational science, in its turn, was to focus on the development and testing of facts in order to arrive at the constitution of an objective representation of educational reality.

An example of this so-called 'positivist' approach is formed by philosophers of education who tried to formulate unequivocal definitions of educational concepts by means of linguistic analysis. The first to introduce this analytic-philosophical method in the Netherlands was Stellwag. Dissatisfied with the lack of unambiguous educational language-use, she tried to clarify a set of concepts that she identified as relevant to all involved in education, yet frequently interpreted variously leading to unnecessary misunderstandings and confusion of tongues (Stellwag, 1970; 1973). Ultimately, Stellwag aimed at formulating universally valid definitions of all foundational educational concepts - to be used by all educationalists, including empiric educational researchers (cf. Meijer, 1997, p. 245).

A related approach can be found in the effort of philosophers of education to contribute to the analysis and constitution of logically consistent systems of claims concerning education, and of the foundational statements in which these systems were grounded (see, for example, Brezinka, 1972; and Stellwag, 1962; 1966). In this respect Stellwag identified distinct 'educational philosophies' that varied according to the ideological point of departure (cf. Heyting, 2006, p. 136). She considered it the task of the philosopher of education to elaborate and clarify such ideologically anchored philosophies: "According to Stellwag, philosophers of education should temper their ambitions, and restrict themselves to mutually comparing those ideological systems and their derived educational prescriptions, and to working out whether they met conceptual and logic standards" (Heyting, 2006, p. 136).

In both approaches outlined above, the philosopher is expected to provide clarity by eliminating on the one hand conceptual ambiguities and, on the other, logical inconsistencies. Clearly, conceptual matters are at the very heart of this analytic philosophy of education. This by no means implies that questions concerning the values of education were ignored but, rather, that any philosophical engagement with such questions needed to be localized at the level of language. In practice, analytic engagement with questions of value took on different forms. When it came to, for instance, education, Peters considered it futile to attempt to make clear a distinction between conceptual issues and questions of values, because the concepts involved - first and foremost that of 'education' itself - are inherently value-laden (see Peters, 1967). In his view, the formulation of rules for the correct use of a concept such as 'education' immediately necessitates an understanding of what is considered valuable in education (cf. Meijer, 1983, p. 325).

Whilst other philosophers of education - such as Soltis and Langford - subscribed to Peters' idea that educational concepts are inevitably value-laden, they argued that philosophical analysis should restrict itself to the neutral identification of the normative moment in education. Such an analysis would help practitioners to make explicit and

14

deliberate normative educational choices (cf. Meijer, 1983, p. 331). A further task for analytic philosophers was the formulation of rules for the correct use of the concepts involved in such processes of normative deliberation. Examples of this form of conceptual analysis are Dutch philosophers of education Spiecker's (1991) and Steutel's (1992) studies of language use regarding, respectively, moral emotions (such as guilt, love, and care), and virtues (such as self-control, justice, and reliability); contributions to a research project on 'moral education'. Finally, analytic engagement with values in education has taken the form of logical analysis of decision-making processes, in which justification of educational claims is sought in supposedly deeper-lying, normative, basis assumptions (cf. Peters, 1974). The issue, then, is not to determine the value of such assumptions, but to evaluate the logical consistency of the process of justification. These analyses are not located at the level of distinguished concepts or claims, but at the level of systems of claims and how. within these systems, claims are mutually interrelated.

The idea that philosophy should confine itself to the level of language is not exclusive to positivists. Looking back, this notion proves to be characteristic for a broad meta-philosophical tradition that endures to this day. Following Bergmann, Rorty (1967) refers to a 'linguistic turn' in the history of philosophy that marked the beginning of a so-called 'linguistic philosophy', characterized by the view that "philosophical problems are problems which may be solved (or dissolved) either by reforming language, or by understanding more about the language we presently use" (Rorty, 1967. p. 3). The huge significance of this meta-philosophical revolution for philosophy of education becomes apparent when we look at the important, perhaps even dominant, role that analytic philosophical approaches have played in international philosophy of education over the past decades.

From the very first, the positivist way of thinking met a great deal of resistance, especially from philosophers of education whose interpretation of the tasks and methods of philosophy of education came closer to the humanities (Geisteswissenschaften) than the natural sciences. Whilst concurring that philosophy of education should primarily deal with the study of educational language-use and questions concerning values in education, these authors did not subscribe to the positivist distinction between philosophy of education and empirical educational science, whereby the former should focus exclusively on analysis and the latter on describing the 'facts' of education.

Their main objection concerned the positivist idea that it was possible to objectively describe and explain educational reality - in the same way as natural reality. These philosophers of education regarded education as an essentially human reality that cannot be adequately studied from the outside. Rather, the study of education should be a matter of understanding it from within (Verstehen); in its totality, as it appears to us in our actually lived experience - its inherent values included. Moreover, since the interpretation of educational phenomena necessary to achieve such an understanding clearly requires a hermeneutic-philosophical engagement with the reality of education, philosophy of education could not confine itself to the analysis of educational language-use.

Furthermore, following Dilthey, these hermeneutically oriented philosophers of education pointed out that any description - and most certainly descriptions of an essentially human reality as education - needed to be understood as a historically anchored (and hence not objective) representation, implying that even 'empirical facts' could not possibly be separated from the contingent historical perspective from which they are formulated. Consequently, empirical claims were also regarded as essentially normatively colored, making it impossible to sharply distinguish these from purely conceptual claims.

These ideas not only affected hermeneutically oriented philosophers of education's attitude towards empirical educational science, they also had an impact on the authors' views regarding engagement with values. The hermeneutic approach, after all, required that values, too, were studied and clarified as historical contingent phenomena. It was no longer considered meaningful to reflect on values separated from the actual historical context in which they are formulated. The study of values in education thus necessitates engagement with history.

The epistemological insights put forward by the hermeneutic school in philosophy of education were not lost on representatives of the positivist approach to educational science. Along the way, they lost some of their initial rigidity. The notion that the empirical acquisition of knowledge, too, was inevitably grounded in a specific theoretical - and hence conceptual - framework, necessitated empirical scientists to abandon their claim of contributing to the construction of an objective picture of external reality. A new understanding of empirical science as a concept-driven activity was now required. Empirical scientists had to accept that, henceforth, they could do little more than formulate hypotheses on how reality might be and, at best, identify false hypotheses to be rejected (cf. Popper). Similarly, analytic philosophy could no longer hold on to its concept of language as an independent, objective apparatus that, used correctly, might help eradicate philosophical problems altogether. In this respect, philosophers like the latter Wittgenstein and Ryle, who conceived of language as conventional, as an instrument as well as a product of (inter-)human activity, proposed that linguistic analysis could ultimately

16

contribute little more than the clarification of current language-use in hopes of eliminating potential misunderstandings; analytically confined to the temporal and spatial context of actual language-use.

The growing reserve within analytic philosophy of education is exemplified by Peters and Hirst. Although they ascribed a categorical and universal status to the descriptions of 'education' (Peters, 1967) and 'liberal education' (Hirst, 1974) they developed early in their careers, they later came to interpret the significance of these educational concepts as culture-specific (albeit, in their view, not completely devoid of objectivity) (cf. Winch, 2006, p. 58). Consequently, both Peters and Hirst abandoned their initial aim of formulating the necessary conditions for any possible interpretation of education. As put by Winch: "[they] re-oriented their view of philosophical analysis to a modest, but more achievable goal of providing *a* conceptual framework for thinking about a central human institution such as education [italics added]" (Winch, 2006, p. 58).

Developments in the area of epistemology, then, appear to have compelled philosophers of education to temper their pretences. Both the analytic and the hermeneutically oriented camp resigned themselves to the fact that philosophy of education could only offer local and temporal descriptions of, either, current educational language use, or of an historically situated educational reality. It seemed as if philosophy of education would have to relinquish its traditional ambition to formulate and justify general and universally valid educational prescriptions once and for all.

It was the so-called '*Kritische Paedagogik*' (critical pedagogy) of the nineteen seventies and eighties, with its roots in the *Kritische Theorie* of the *Frankfurter Schule*, that responded to the growing reticence over the feasibility of a prescriptive role for philosophy of education. Repre-sentatives of the Frankfurt school of philosophical and social thought, such as Habermas, had pointed out that current approaches to academic endeavor ignored the fact that the acquisition of knowledge was not only situated in spatial and temporal contexts, but also driven by societal conflict, personal and/or group interests and dispute (cf. Kunneman, 1986, p. 194). Against this background, critical theorists argued, that academics could not turn their back on their social responsibilities. If research and theory were not only incapable of achieving objectivity, but also plainly partial, and in that sense political, then one was socially and morally obligated to go beyond mere description. Academics were to be held accountable for their influence on society, and required to contribute consciously to societal transformation.

Critical pedagogy aligned with the hermeneutically oriented philosophy of education in its rejection of (neo-)positivism in the human sciences. In these authors' view, the positivist approach wrongly maintained that the scientific methods of the natural sciences were the only objective and hence only valid methods for research. Any form of research driven by a specific knowledge-governing interest and could, therefore, not possibly be objective. In the case of (neo-)positivism, these authors contended, (scientific) research was driven by technical interests aimed at controlling human behavior, resulting in a specific, politically colored, description of educational reality. Besides epistemological, practical objections were also advanced. Since positivists interpreted human behavior solely in terms of causes and effects, human actions - for instance, in education - were reduced to a purely instrumental matter, thus dehumanizing society.

From a critical pedagogical perspective, hermeneutically oriented educational science had a more humane outlook on education because its interpretative claims did not aim to predict and control human action, but to conduct a social dialogue within the human sciences; a dialogue that acknowledged individual human experiences, motives and intentions. However, critical pedagogues did not consider the hermeneutic approach to the human sciences flawless. Because of the descriptive nature of its knowledge-claims, a hermeneutic approach to educational research was not able to critically examine the social injustices that characterized actual social communication, Critical pedagogues argued. Hermeneutic educational research was thought to be governed by the practical interests existent in everyday - historically and culturally situated - educational practice, as a result of which the knowledge it produced was too affirmative in character. Consequently, hermeneutic research could only reproduce social realities, and would never be able to contribute to societal transformation. In addition, hermeneutically oriented philosophy of education was considered too philosophical and too speculative, underestimating the value of rigorous empiric-analytic educational research.

Critical pedagogy wanted to show that existing approaches to educational science had been blind to the idea that any form of academic endeavour served specific interests and neglected others, and were thus oblivious to the ideological nature of academic enquiry. Critical pedagogy's main focus was that the acquisition of knowledge was inevitably involved in the construction and/or reproduction of a specific societal structure, and hence also in the creation and/or preservation of the unequal distribution of power and resources within societies (cf. Kunneman, 1986, p. 230-231). Against this background, and in the light of the critical pedagogic normative point of departure that all societal institutions ought to serve emancipatory interest, these authors held that philosophy of education had a responsibility to (strive to) influence and transform societal processes that instigated or reproduced forms of social injustice. In this, one of the tools philosophy of education had at its disposal was the critical analysis of existing societal structures and their underlying mechanisms of power - critique of ideology.

Apparently, critical pedagogy did not aspire to formulate 'objective' judgments of (social) reality. Rather, it endeavored to formulate a philosophically justifiable way of judgment-making that was resistant to the power-mechanisms underlying current society. It was Habermas (1981) who was inspirational to many authors on this point. He strived for a form of consensus that he thought could be reached on the basis of a dominion-free dialogue; a dialogue driven, ultimately, by emancipatory interests. Such a dialogue should embrace empiric-analytic, hermeneutic-interpretative. and ideological-critical knowledge claims, as long as none of these dominated the dialogue (cf. Miedema, 1997. p. 132-133). With its plea for a dominion-free dialogue, and the critical role it could play within such dialogue, critical pedagogy regained the prescriptive ambitions that had been so characteristic for philosophy of education in its early years.

The idea of a dominion-free dialogue can be regarded as an attempt to break free from the cultural and historical situatedness - brought to our attention by the Geisteswissenschaften - and interest-governed perspectivity - identified by the Kritische Theorie - of any claim. It was a methodological point of departure that would lead to - in a philosophical sense - universally justified claims. In that respect it resembled Rawls' idea of the so-called 'original position'; a point of view not contaminated by social interests, that could serve as a starting point for designing a model for a just society (Rawls, 1971).

Rawls' idea of the 'original position' has been of major importance for analytic philosophy of education from the nineteen seventies to the present day. The 'original position' is thought- experiment that requires the participant to think of a hypothetically impartial position in society from which the rules for a just society should be drawn up. According to Rawls, the 'original position' had to meet two criteria. Firstly, the participant should only want to serve his/her own interests in the way most beneficial to him/herself. Secondly, this participant would be placed behind a so-called 'veil of ignorance', meaning that he/she would have no knowledge whatsoever of the characteristics and interests of the person or group in society that he/she represents (d'Agostini, 2003). According to Rawls, this thought-experiment enables the formulation of ethico-political claims whereby the criterion of egocentrism rules out the possibility of sacrificing the interests of individuals on behalf of those of the majority, whilst the 'veil of ignorance' ensures that no specific persons or groups are favored over others. Rules drawn up from the 'original position' were therefore supposed to be 'just', by definition. To give an example: a person in the 'original position' would have no knowledge of his or her sex. This, in combination with the person's egocentric nature, would keep him or her from drawing up rules that would privilege the interests of either women or men. At least for as far as equality of the sexes would be concerned, the rules drawn up by such a person would have to be 'just'.

Rawls tried to break loose from the partiality of human judgment-making, just as Habermas did. Both thought to have found the solution in a specific procedure that would enable the formulation of philosophically justified claims without any form of partiality. Habermas and Rawls provided tools for philosophers of education to return to their initial prescriptive ambitions; not, now, by taking a normative stance, but by carefully following a supposedly politically neutral procedure. However, as philosophers of education realized only later, these procedures were, in themselves, grounded in conventional and hence value-laden concepts (such as 'emancipation' and 'justice'), thus privileging specific social interests in their turn.

From the nineteen eighty's onwards, many philosophers of education continued to hold on to the idea that they should engage with conceptual questions as well as with questions concerning value. However, they had to give up all hopes of ever being able to formulate general and final answers to such questions (for instance, regarding conceptions of 'the good life', or the correct use of educational concepts) in a scientifically justified way. The idea of being philosophically impotent instigated a broad adherence to the thought that we are confronted and left with an incommensurable diversity of 'conceptions of the good life' or conceptual frameworks. According to Rorty, philosophers had to recognize that the fruit of their labor inevitably gives expression to a specific local value-laden point of view - in Rorty's case a liberal perspective. This recognition, Rorty argued, would inevitably lead us to a position in which that very point of view is put into perspective, and in which the existence and potential value of alternative viewpoints is acknowledged; a paradoxical position that he refers to as 'anti-anti-ethnocentrism' (Rorty, 1991, pp. 203 ff.).

Some philosophers of education went even further; not only acknowledging the incommensurable diversity philosophy was left with, but even considering it their task to explicate and celebrate this diversity. An example can be found in Rang's plea for different forms of pluralism; a concept that he defines as an appreciative attitude towards a controversial form of diversity - for instance in the areas of morality, science, art, religion, or culture (cf. Rang, 1993, p. 19-21). In the nineteen eighties and nineteen nineties, such an attitude was apparent across various academic domains. In philosophy of education it was expressed by authors who can be associated with epithets such as: 'feminism' (among others Noddings, 1984; Roland Martin, 1985; and Nicholson, 1990); 'postmodernism' (among others Usher & Edwards, 1994; and Aranowitz & Giroux, 1991); and 'multiculturalism' (among others Feinberg, 1996).

At the same time, however, the idea that we are left with an incommensurable diversity of viewpoints has been contested from the start (see among others Carr, 1998; and Siegel, 1997). According to Siegel (1997), this idea is epistemologically incorrect, because it relies on the rejection of any claim to universal validity. Siegel starts by pointing out that such a rejection is logically inconsistent, because it presupposes a claim to universality validity itself - i.e. the claim that any claim to universal validity would be invalid (p. 175). But that is not his only coun-terargument. In his mind an even stronger objection can be made to the substantiation of the idea that any claim to universality is philosophically incorrect by arguing that our judgments are inevitably bound by a specific - and in that sense restricted and excluding - conceptual framework. Siegel posits that there are numerous examples of claims (for instance in mathematics or physics) of which the validity and correctness extend far beyond the boundaries of the frameworks in which they were developed or made. The law of universal gravitation, one might argue in this respect, also applied to Francis Bacon - even though he did not share the conceptual perspective of Newton, who first formulated the law (Bacon died seventeen years before Newton was born). According to Siegel, this shows there is no reason to think that the acceptance of an incommensurable diversity of particular perspectives is unavoidable, and that we should end our search for universal knowledge claims.

Besides epistemological objections, more practical objections have more recently been uttered. Some philosophers have warned that embracing pluralism can only result in "a rampant relativism, leading to nihilism and social anomie" (Kvale, 1992, p. 8, cited in Usher & Edwards, 1994, p. 26). The major concern is that an ongoing focus on diversity puts social cohesion under pressure. In this view, 'pluralism' is merely a fancy name, used to give a positive twist to processes of social disintegration. Once again, there is a call for direction and unity. Especially after the terrorist attacks on New York of September eleventh 2001, there seems to be a growing fear that liberal society as we know it will fall apart, as social groups are seen to become increasingly and ever more diametrically opposed to each other. The result is a growing demand for factors that might serve to bind all members of society together.

The search for something that might unite all people within liberal societies - societies that by definition, to a certain extent, accept and even appreciate various forms of diversity - is also evident in the work of many present-day philosophers of education. (see f.i. Callan, 1997; and Burtonwood, 2006). They are motivated by the idea that, in order to keep social structures from collapsing, all individuals within those structures need to share some common ground. Against this background, Spiecker en Steutel (1995; 2003), for instance, investigate which common values need to be maintained by, or even imposed upon, members of pluralist societies in order to protect social unity.

Regardless of the way these objections to pluralism are valued or interpreted, the recent call for unity unmasks pluralism as another value-laden and thus disputable point of departure. The thought that any possible starting-point is arbitrary increasingly forces itself upon us, (one again) bringing philosophy of education's tasks and methods into question. The constantly evolving discussion, set in with the 'linguistic turn', over the issues philosophy of education should address, and how these issues should be addressed, now seems to be becoming ever more divergent. Against this background, it seems that we need to abandon the hope of ever being able to develop an unequivocal understanding of the tasks and possibilities of philosophy of education as an academic discipline.

3. Research-questions

In the remainder of this dissertation, I will deal with the question of what may yet be expected of a future philosophy of education. What tasks can be assigned to philosophy of education? What role will philosophy of education, considering the (epistemological) issues it now appears to be confronted with, be able to play within the broader field of educational science? And, what is its relation to educational practice? The questions to be dealt with in the following chapters have been formulated against the background of these central questions. Because the answers to the questions in each chapter provide the backdrop to the questions raised in the consecutive chapter, this dissertation can be read as a continuous story. However, chapters two, four, six, and eight were written as independent research-articles, each dealing with their own research-questions, and, hence, they can also be read independently from the other chapters.

In the next chapter (chapter two) I investigate how present-day philosophers of education have responded to the rejection of the foundationalist model of justification, that is considered to be at the heart of the radical epistemological doubts expressed by a vast group of philosophers in the past decades. It turns out these so-called antifoundationalist philosophers of education do not interpret such episte-mological issues and the undecidability - i.e. the impossibility of ever being able to formulate final answers to educational questions - these imply as an intellectual obstacle, but rather as an opportunity to formulate an alternative outlook on the tasks of philosophy of education. Throughout history, the continuous reinterpretation of philosophy of education's central tasks - partly instigated by developments in epistemology - has always been related to changing ideas about the discipline's practical relevance. Against this background, chapter two will also investigate what the authors included in this study present as the practical relevance of philosophy of education in light of their renewed interpretations of its central tasks.

The antifoundationalist approach reconstructed in chapter two is, of course, not the only possible response for philosophy of education facing epistemic uncertainty. Moreover, it has been criticized from the very beginning. In order for me to be able to develop my own position in the debate, in chapter three I look more closely into the epistemological arguments behind the idea of an inescapable epistemic uncertainty. Consequently, I elaborate on an epistemological approach defended by a group of philosophers of education that accept the idea that knowledge will always be - to a certain extent - uncertain, but that respond to this idea in more moderate way than the antifoundationalist philosophers of education examined in chapter two. I examine the tenability of this position. In the light of this examination, and the conclusions of chapter two, I conclude that it seems fruitful to further examine contextualism as a possible acceptable epistemological position that enables us to deal with the idea epistemic uncertainty. The question is raised what may still be expected from a philosophy of education incorporating a contextualist epistemological position.

In chapter four I explore one way for a philosophy of education that takes a contextualist approach towards epistemic uncertainty to still contribute to educational thinking in a constructive and meaningful manner. In this chapter, the ultimate undecidability regarding educational questions, that is implied with the idea of epistemic uncertainty, forms the starting point. I systematically investigate how this undecidability can be dealt with in a philosophically acceptable, and at the same time meaningful way, through the use of irony as a philosophical tool. In response to two recent ironic approaches in general philosophy, I develop my own ironic approach, based on the idea of dynamic contexts of justification. At the end of the chapter the three different approaches to a so-called 'ironic philosophy' are presented and the contributions that may be expected of these approaches are illustrated with reference to debates around the educational issue of 'students at risk'.

Chapter four yields ideas concerning how human meaning-making, inter-human communication, and the relation between language and reality can be understood in light of a dynamic-discursive interpretation of contexts of justification. These ideas are relevant to the overarching aim of this dissertation. In chapter five I examine how these ideas may contribute to reaching that aim, i.e.: developing an understanding of epistemic (un)certainty and considering the consequences of such an understanding for (the tasks and possibilities of) philosophy of education. At first, I show that using a dynamic-discursive interpretation of context enables me to repudiate the reproach of relativism that contextualist approaches to epistemic uncertainty usually seem to evoke. Secondly, based on the idea of dynamic-discursive contexts of justification, I sketch the contours of my own discursive interpretation of epistemic uncertainty. However promising the developed epistemological approach may be, at the end of the chapter I argue that it is vulnerable to the reproach of being too conventionalist, which instigates me to further examine the ideas behind the epistemological approach.

In chapter six these ideas will be explored in greater depth by applying them to the debate concerning 'differential academic language proficiency in schools'. The question is raised what it means for a human subject to learn (to use) a language. Consecutively, I investigate what inferences can be drawn from the answer for dealing with differences in academic language proficiency. It appears that a subject learning to use a language may best be understood as an active participant in an ongoing negotiation concerning how the world is, which implies a specific interpretation of differential academic language proficiency that may also appeal to a different way of dealing with this issue in, for instance, educational policy.

Chapter six gives us an insight into how the active role of participants in communicative processes can be understood. It shows how communicative contexts are transformed with every – successful – contribution to the communication made by a speaker. In chapter seven, these insights are used to show that a discursive epistemology cannot be regarded as conventionalist, at least not in any conservative sense. Furthermore the consequences of such an understanding of evolving communicative contexts a drawn for an idea of what it means to speak of the 'growth of knowledge' in academic disciplines. In light of the findings, it is argued that understood from the perspective of a discursive epistemology the commitment of individual academics plays an important role in the development of knowledge within academic disciplines, leading me to conclude that 'commitment' may also have re-entered philosophy of education.

Making use of the insights developed in the consecutive chapters, in chapter eight I investigate how the concept of 'commitment' should, in this respect, be interpreted, how an apparently committed philosophy of education may be understood, and what might be expected of such a philosophy of education.

In the concluding chapter (chapter nine) I shall begin by summarizing the successive findings from the preceding chapters. Subsequently, I will draw out the conclusions concerning what I have come to see as a philosophically acceptable, and fruitful epistemological position for philosophy of education, and I will elaborate on the possible consequences of such a position for an understanding of the tasks and possibilities of a future philosophy of education and its relation to educational practice.

4. References

Aranowitz, S. & Giroux, H. (1991). *Postmodern education.* Minneapolis: University of Minnesota Press.

Blake, N., Smeyers, P., Smith, R., & Standish, P. (1998). *Thinking again. Education after postmodernism.* London: Bergin & Garvey.

Brezinka, W. (1972). *Von der Paedagogik zur Erziehungswissenschaft. eine Einführung in die Metatheorie der Erziehung.* Weinheim: Betz.

Burtonwood, N. (2006). Cultural *diversity, liberal pluralism, and schools. Isaiah Berlin and education.* London: Routledge.

Callan, E. (1997). *Creating citizens: political education and liberal democracy.* Oxford: Clarendon Press.

Carr, D. (Ed.)(1998). *Education, knowledge, and the truth: beyond the postmodern impasse.* London: Routledge.

D'Agostini, F. (2003). Original position. In Zalta E. N., Nudelman, U. & Allen, C. (Eds.), *Stanford encyclopedia of philosophy.* Retrieved Septembre 24, 2008, from http://plato.stanford.edu/entries/original-position/

Feinberg, W. (1998). *Common schools uncommon identities: National unity and cultural differences.* New Haven: Yale University Press.

Habermas, J. (1981). *Theorie des kommunikativen Handelns.* Frankfurt am Main: Suhrkamp

Heyting, F. (2001). Methodological traditions in philosophy of education: introduction. In F. Heyting, D. Lenzen & J. White (Eds.), *Methods in philosophy of education* (pp. 1-12). London: Routledge.

Heyting, F. (2006). The contingency of scholarly development across generations. In R. Van Goor & E. Mulder (Eds.), *Grey wisdom? Philosophical reflections on conformity and opposition between generations* (pp. 131-137). Amsterdam: Amsterdam University Press.

Hirst, P. H. (1974). *Knowledge and the curriculum.* London: Routledge.

IJzendoorn, M. H. van (1997). Empirisch-analytische pedagogiek. In S. Miedema (Ed.), *Pedagogiek in meervoud. Wegen in het denken over opvoeding en onderwijs* (pp. 73-116). Houten: Bohn Stafleu Van Loghem.

IJzendoorn, M. H. van (2002). Methodologie: kennis door veranderen, de empirische benadering in de pedagogiek. In M. H. van IJzendoorn & H. de Frankrijker (Eds.), *Pedagogiek in beeld. Een inleiding in de pedagogische studie van opvoeding, onderwijs en hulpverlening* (pp. 17-31). Houten: Bohn Stafleu Van Loghem.

Kunneman, H. (1986). *De waarheidstrechter. Een communicatietheoretisch perspectief op wetenschap en samenleving.* Amsterdam: Universiteit van Amsterdam.

Kvale, S. (Ed.) (1992). *Psychologie and postmodernism.* London: Sage Publications.

Levering, B. (2003). From schools of thinking to genres of writing. In P. Smeyers & M. Depaepe (Eds.), *Beyond Empiricism. On criteria for educational research* (pp. 93-103). Leuven: Leuven University Press.

Lyotard, J.-F. (1984). *The postmodern condition: A report on knowledge* (G. Bennington & B. Massumi, Trans.). Manchester: Manchester University Press.

Meijer, W.A.J. (1983). Angelsaksische analytische filosofie van opvoeding en onderwijs. In J. D. Imelman (Ed.), *Filosofie van opvoeding en onderwijs. Recente ontwikkelingen binnen de wijsgerige pedagogiek* (pp. 321-339). Groningen: Wolters-Noordhoff.

Meijer, W. A. J. (1997). Taalanalytische pedagogiek. In S. Miedema (Ed.), *Pedagogiek in meervoud Wegen in het denken over opvoeding en onderwijs* (pp. 215- 261). Houten: Bohn Stafleu Van Loghum.

Miedema, S. (1997). Kritische pedagogiek. In S. Miedema (Ed.), *Pedagogiek in meervoud Wegen in het denken over opvoeding en onderwijs* (pp. 117-169). Houten: Bohn Stafleu Van Loghum.

Mulder, E. (1989). *Beginsel en beroep. Pedagogiek aan de universiteit in Nederland 1900-1940.* Amsterdam: Universiteit van Amsterdam.

Nicholson, C. (1990). Postmodernism, Feminism, and Education: The Need for Solidarity. *Educational Theory, 40*(1), 197-205.

Noddings, N. (1984). *Caring: A feminist approach to ethics and moral education.* Berkeley: University of California Press.

Roland Martin, J. (1985). *Reclaiming a conversation: the ideal of the educated woman.* New Haven: Yale University Press.

Peters, M. (1995). Introduction: Lyotard, Education, and the postmodern condition. In M. Peters (Ed.), *Education and the postmodern condition* (pp. xxix-xlix). London: Bergin & Garvey.

Peters, M. & Lankshear, C. (1996). Postmodern counternararatives. In H.

Giroux, C. Lankshear, P. McLaren & M. Peters (Eds.), *Couternarratives. Cultural studies and critical pedagogies in postmodern spaces*. London: Routledge.

Peters, R. S. (1967). *The concept of education*. London: Routledge.

Peters, R. S. (1974). *Ethics and education*. London: Routledge.

Rang, A. (1994). Padagogik und Pluralismus. In F. Heyting & H. E. Tenorth (Hrsg.), *Pädagogik und Pluralismus. Deutsche und Niederlandische Erfahrungen in Umgang mit Pluralität in Erziehung und Erziehungswissenschaft*. Weinheim: Deutscher Studien Verlag.

Rawls, J. (1971). *A theory of justice*. Cambridge: Harvard University Press.

Rorty, R. (1989). *Contingency, irony, and solidarity*. New York: Cambridge University Press.

Rorty, R. (1991*). Objectivity, relativism, and truth. Philosophical papers, volume 1*. New York: Cambridge University Press.

Rorty, R. (1967). Introduction. In R. Rorty (Ed.), *The linguistic turn. Recent essays in philosophical method* (pp. 1-39). Chicago: The University of Chicago Press.

Siegel, H. (1997). *Rationality redeemed? Further dialogues on an educational ideal*. New York: Routledge.

Slife, B. D. & Williams, R. N. (1995). *What's behind the research? Discovering hidden assumptions in the behavioural sciences*. London: Sage Publications.

Spiecker, B. (1991). *Emoties en morele opvoeding. Wijsgerig-pedagogische studies*. Amsterdam: Boom.

Spiecker, B. & Steutel, J. (1995). Political liberalism, civic education and the Dutch government. *Journal of Moral Education. 24*(4), 383-394.

Spiecker, B. & Steutel, J. (2003). Zelfconcept en maatschappelijke integratie. *Pedagogiek, 23*(4), 318-329.

Stellwag, H. W. F. (1962). Positief. of negatief. *Pedagogische Studiën, 39*(7/8). 321-332.

Stellwag, H. W. F. (1966). De verhouding van de wetenschap der opvoeding tot de praktijk. *Pedagogische Studiën, 43*, 97-119.

Stellwag, H. W. F. (1970). *'Situatie' en 'relatie'*. Groningen: Wolters-Noordhoff.

Stellwag, H. W. F. (1973). *'Gezag' en 'autoriteit'*. Groningen: H. D. Tjeenk Willink.

Steutel. J. W. (1992). *Deugden en morele opvoeding*. Een wijsgerig-pedagogische studie. Amsterdam: Boom.

Usher, R. & Edwards, R. (1994). *Postmodernism and education*. London: Routledge.

Waterink, J. (1959). Enkele vraagstukken betreffende de paedagogiek als wetenschap. In R. L. Plancke, L. Coetsier, W. De Coster, P. De Keyser en R. Verbist (Eds.). *Naar een verantwoorde opvoeding: album Prof J. E. Verheyen* (pp. 188-192). Gent: Rijksuniversiteit Gent.

Winch, C. (2006). On the shoulders of giants. In R. Van Goor & E. Mulder (Eds.). *Grey wisdom? Philosophical reflections on conformity and opposition between generations* (pp. 53-74). Amsterdam: Amsterdam University Press.

II
BEYOND FOUNDATIONS - SIGNS OF A NEW NORMATIVITY IN PHILOSOPHY OF EDUCATION[2]

1. Introduction

How philosophy of education can contribute to practical educational thinking partly depends on available methodological resources. Consequently, any revision of established philosophical approaches might require reformulation of the practical relevance one attaches to philosophy of education. Philosophical developments of the past decades seem to affect one of the main tasks traditionally attributed to philosophy of education: that it should contribute to the formulation and justification of the founding principles of education (Snik et al, 1994). However, accomplishing this task presupposes a model of philosophical justification that has been under discussion for several decades now.

To illustrate this model of justification, I take an example from De Ruyter (2003). She claims the importance of passing on ideals in education (Y), and substantiates this claim with the proposition that ideals are indispensable guides for finding meaning in life (X). In other words, she justifies her claim (Y) by reducing it to another claim (X) that is presented as a (more) reliable one. De Ruyter departs from an anthropological foundation, which ascribes specific essential characteristics to human nature that are presented as undisputed and therefore reliable. Because of this alleged reliability, the latter claim (X) - attributing the need for secure guidance and meaning to mankind - can function as a foundation for ensuring the reliability of derived claims (such as Y). Most philosophers of education are reticent about how certain a foundation should be in order to deserve this status. They believe that no basic foundation (X) can be considered irrefutable, as a second example can illustrate. Snik argues that if one accepts autonomy-based values as a basic principle for evaluating state interference in education (X'), it can function as a foundation for justifying derived claims such as the claim that any school should respect the child's right to develop into an autonomous human being (Y') (Snik, 1999). Like De Ruyter, Snik relates the validity of conclusion Y' to a more fundamental principle (X'). In contrast with De Ruyter, however, Snik explicitly puts the epistemological status of his foundation (X') into perspective - that is, he presents the

2 Published as: Van Goor, R., Heyting, F. & Vreeke, G.-J. (2004). Beyond foundations. Signs of a new normativity in philosophty of education. *Educational Theory, 54*(2), 173-192 (printed with permission)

acceptability of his conclusion as conditional, dependent on the acceptance of his basic principle.

Still, the model of justification is the same in both examples. According to this model, justification of (educational) claims hinges on (the acceptance of) more certain propositions. In addition, this model implies the acceptance of specific rules with respect to 'correct' procedures for reducing claims to foundations in order to justify them. In other words, according to this model, justification depends on accepted foundations as well as accepted (usually called 'rational') forms of argumentation.

Owing to recent developments in epistemology, both characteristics of this justification model are presently under attack (Dancy, 1985; Rorty, 1979). This may lead to changes in methodology that, as mentioned previously, could influence ideas about the main tasks and relevance of philosophy of education. Against this background, I decided to investigate how philosophers of education responded to epistemological discussions of this model of justification and how their views about the model affected their positions regarding the tasks and practical contributions of philosophy of education. To this end, I evaluated publications from the field of philosophy of education that both explicitly reject this model of justification and that suggest an alternative approach. I reviewed a range of materials, including books, book chapters, and journal articles published since 1995, when the subject started to receive substantial attention in philosophy of education[3].

In the next section, I will explain the disputed model of justification in more detail. Then I will present the results of my textual analyses, arranged according to the following research questions: What epistemological objections against the justification model do these philosophers emphasize most when they reject it? What loss of relevance for philosophy of education do they notice as a consequence of this rejection? What alternative views regarding justification do these authors suggest? How do specific alternative approaches affect the relevance of philosophy of education according to these authors? Generally speaking, the results of my analysis show that these philosophers of education are inclined to replace foundationalist views of justification with contextualist alternatives. In a concluding section, I will discuss whether such alternatives are tenable in the light of some critical objections.

---

3 In our survey of the literature, we included articles from the following journals in the field: *Educational Theory*; *Educational Philosophy and Theory*; *Journal of Philosophy of Education*; and *Studies in Philosophy and Education*.

2. Foundationalism

The term 'foundationalism' refers to a model of justification in which a claim's acceptability is considered to depend on whether it is possible to rationally reduce it to a more certain claim (or claims) that then functions as a compelling argument (or arguments) in support of the original claim. In principle, from a foundationalist point of view assertions can be divided into two groups. The first group includes those assertions that are justified only by other assertions. The second group includes assertions that are considered justified independent of other assertions, and in that regard this second group enjoys a certain epistemological superiority. By virtue of this status, they can serve as basic foundational grounds for the justification of other assertions. Assertions from the first group can only serve as grounds for justification insofar as they can be justified in their turn, and so on, until one arrives at an assertion from the second group, which is considered independently justified. Thus, the foundationalist model of justification arranges assertions in a hierarchical structure, and this structure implies that foundations, or 'last grounds', constitute the end of the sequence. In 'classical' foundationalism these foundations are considered self-justificatory because of their special epistemological characteristics (Dancy, 1985, p. 53-4). Rorty discusses three varieties of classical foundationalism - empiricism, rationalism, and transcendentalism - each characterized by its specific brand of epistemological privilege (Rorty, 1979).

Since the Vienna Circle, the idea of self-justificatory foundations has been called into question, and most philosophers today reject such a notion. However, there still is substantial support for another kind of foundationalism, in which the hierarchical structure of justification is preserved but foundations are no longer considered infallible. Dancy refers to this as 'weak foundationalism' (Dancy, 1985, p. 62). In 'weak' foundationalist approaches justification consists of reducing claims to 'final grounds' that are not considered infallible but that are still endowed with relative epistemological privilege. They are treated as if they were more certain - and thus less in need of justification - than other claims. In this sense a "special - though not absolute - relation to truth" is still attributed to 'foundations', and it is on this basis that they can be used to justify other claims (Heyting, 2001, p. 110). A good example of this hierarchical, weak foundationalist model of justification is Snik's justification of education system's duty to respect the child's right to develop into an autonomous human being (Snik, 1999). He does not consider the claims of an autonomy-based perspective on liberal morality, which serve as foundations for his argument, self-justifying. He notes that his conclusions will only be valid for those who endorse these values. However, by being endorsed, these values seem to gain a relative epistemological privilege that makes them suitable for the justification of other claims.

Foundationalism not only implies a specific justificatory structure; it also implies the necessity to acknowledge specific rules of inference. Only claims that were 'correctly' reduced to foundations can be accepted as justified. Mere association, for example, is not enough. Usually, established rules of 'rational' argumentation must have been applied for a justification to be considered convincing. In such cases. the involved principles of rationality gain an epistemological status that is comparable to that of (classical or weak) foundations. For example, in Habermas's theory, any concluding consensus that is reached according to the 'right' principles of discourse can be considered justified. These procedure-oriented varieties of the foundationalist model can be viewed as either 'classical' or -weak'. In the philosophy of language, which strongly influenced British philosophy of education (see, for example, the work of Peters, Dearden, and Hirst), we can distinguish similar positions with respect to justification. Originally, linguistic philosophers such as Russell, Wittgenstein (in his Tractatus), and Moore were convinced that language analysis could yield knowledge about the structure of reality, because in their view the possibility of linguistically stating a certain fact required that language have something in common with the ontological structure of that fact. Consequently, they expected to find reliable foundations by developing a logically pure language that would make an undistorted representation of reality possible and thus would prevent complicated philosophical problems from arising (cf. Copi. 1967: Smeyers, 2001). After World War II most philosophers (including the later Wittgenstein) rejected this kind of classical foundationalism, and weaker forms of foundationalism started to emerge in linguistic philosophy. Philosophers abandoned the pursuit of a language that would enable a perfect representation of reality. increasingly stressing the conventional nature of any language (Wittgenstein, 2001). Consequently. analytic philosophers now concentrate on describing rules for the correct use of concepts within conventional language games. The conceptual clarification this kind of linguistic analysis promises is not accurate representation of external reality, but only correct usage, as compared to the specific language game in question (White & White, 2001). This approach can be seen as a form of weak foundationalism because the rules for the 'correct use' of concepts are derived from the actual language to which these concepts belong, which gives them only relative certainty.

3. Problems of foundationalism

In the publications I analyzed, authors reject the foundationalist justi-

fication model for a number of different reasons. I distinguish two categories of reasons that correspond with the previously mentioned characteristics of foundationalism: attribution of (relative) epistemological
privilege (1) to basic foundations and (2) to specific rules for correct
argumentation. I then describe what kinds of critique I found with respect
to both elements of foundationalism.

3.1 The problem of establishing epistemologically privileged assertions
A number of publications emphasize the problem of how to establish
epistemological privilege (Biesta, 2001; Giroux, 1997, pp. 147-163 & pp.
183-233; Greene, 1995; Heyting, 2001; Peters, 1995; Peters & Lankshear,
1996). For example, Heyting argues that "neither particulars [obser-
vations] nor universals [methods, theories] can be attributed - relative -
epistemological privilege" (2001, p. 110). Because any attempt to demonstrate such privilege would inevitably result in infinite regression, leaving
us unable to prove any possible 'last' foundation less arbitrary than
another. Maxine Greene stresses, "once we give priority to the signifier
and realize that words refer and relate to other words, not to some
objective world beyond, meanings proliferate and become richer. Hierarchies of meaning, hierarchies in general, become more and more absurd"
(1995, p. 10).

Many authors do not confine themselves to substantiating the
disputable character of attempts to prove epistemological privilege. They
also draw attention to the problem of aiming for knowledge from a so-
called 'view from nowhere', an ideal supported by classical as well as
weak foundationalists. These critics point out that the validity of assertions is inescapably influenced by the personal and social context, and
contrary to foundationalist approaches, they reject the foundationalist
approach of trying to avoid this problem by means of foundations that are
presumed to be (relatively) certain independent of any context. In their
view, foundations inevitably represent specific partial perspectives,
causing all related knowledge to be partial as well. As Child, Williams,
and Birch put it, "if perspectival, then partial. And if partial then not
possibly universal, ultimate, or certain" (1995, p. 167). These authors
primarily draw attention to the idea that knowledge is always embedded in
variable human contexts.

A third group of authors emphasizes this embedded nature of
knowledge, specifically characterizing it as a kind of social embedding
that especially represents (Biesta, 1998; Blake et al, 2000; Child et al,
1995; Fitzsimons & Smith, 2000; Giroux, 1997; Thompson & Gitlin,
1995; Greene, 1995; Gur'ze-ev, 1998; McLaren & Giroux, 1997; Usher et
al, 1997). They often refer to Foucault's notion of 'power/knowledge',

indicating that any knowledge claim will inevitably imply power. Usher, Bryant, and Johnston (1997), for example, point to Foucault's 'regimes of truth', discourses in which establishing knowledge and exerting power go hand in hand Blake, Smeyers, Smith, and Standish explicitly confront traditional hopes that knowledge rests on objective and solid foundations, which would make it not only neutral with respect to power, but also suitable to resist it: "If foundations of knowledge provide the basis on which one might 'speak the truth to power', then the search for them is of paramount importance....Yet the search has failed, on every front: in the theory of meaning, the theory of truth, the theory of knowledge and in philosophy of mind. This might be taken to mean that one cannot speak the truth to power. But on the other hand, it might rather mean not that one cannot do so, but that securing foundations is the wrong preparation for doing so" (1998). In this example, antifoundationalism is directly related to the relevance of philosophy of education, a topic I will address in detail later in this paper.

3.2 The problem of establishing epistemologically privileged procedures
The second aspect of foundationalism concerns the procedures for justifying claims. The texts I reviewed attack this dimension in three ways as well: authors criticize the conceptions of rationality, the notions of correct language use, and the characteristics of the human faculty of cognition that are implicit in these procedures for justifying claims.

Many philosophers argue that there is no such thing as the correct (or even the best possible) standard of rationality. They emphasize the limitations on validity for each procedure of justification (Biesta, 2001; Kohli, 1998; Masschelein, 1998; Peters & Lankshear, 1996; Peters & Marshall, 1999; Ruhloff, 2001; Säfström, 1999; Weinstein, 1995). Reacting to Harvey Siegel's transcendental justification of rationality standards (Siegel, 1987), Weinstein asserts, "All practices, including the philosophical practices that seek to develop general and a priori theories of rationality, are thus essentially limited by particular assumptions, limited points of view, and characteristic procedures of inquiry" (1995, p. 380). Weinstein attempts to demonstrate that Siegel's transcendental argument in favor of the a priori nature of rationality principles already presupposes certain criteria for correct reasoning, such as the meaning and function of 'necessary' and 'sufficient' conditions. According to Weinstein, Siegel's approach is tantamount to withdrawing these criteria from critical scrutiny.

The second critique against privileging certain procedures of justification relates to the use of language. Some authors emphasize that conceiving of language as an unambiguous, representative structure - a conception that is supposed to underlie foundationalist approaches - is untenable (Kohli, 1998; Marshall, 1995; and 1996). This line of argument seems most directly to undermine classical versions of linguistic philosophy that presuppose a representative view of language. Other critics primarily challenge conventionalist-that is, weak foundationalist – versions of linguistic philosophy. For instance, Michael Peters and James Marshall argue that even within language games, one will never be able to define clearly and completely the rules for the 'correct' use of concepts. They emphasize the dynamic and creative nature of language use, which cannot be understood as merely following rules. Insofar as we can speak of 'rules' in the use of language, they are constantly in the making and at best can be reconstructed in retrospect. Furthermore, we cannot attribute any normative meaning to the results of such a reconstruction.

The third reason for rejecting procedural privilege concerns presupposed features of the knowing human subject that are often used as last foundations for justifying justification procedures. This third form of critique can be found in Peters and Marshall as well. They endorse Jean-Francois Lyotard's attack on 'the concept of universal reason and of the unity of both language and the subject', in which all three varieties of critique of procedural privilege are involved and mutually connected. According to Peters and Marshall, rationality standards, language use, or the knowing subject are interrelated because the knowing subject is seen as the bearer of both language and rationality. They point out that rejecting the foundational nature of language implies that one also must abandon any idea of 'the' human subject, because every human being at any moment finds him- or herself at a crossroads of many different language games. Peters and Marshall further note that human beings actively contribute to the development of these language games (1999, p. 126). Peters and Lankshear's (1996) argumentation runs along similar lines.

Some authors attack conceptions of the knowing subject as a basis for procedural privilege from a different perspective. They concentrate on the presupposed portrayal of humankind in foundationalist approaches. Denise Egea-Kuehne (1995), for example, criticizes the assumption that the 'conscious self' is at the center of all human activity. Child, Williams, and Birch (1995) also reject the 'first-person perspective' assumed in the traditional model of justification, arguing that justification should not be based on ontology of the knowing human individual. They see an intrinsic connection between ethics and justification. much like the previously mentioned group of philosophers intrinsically relate knowledge and power. Following Levinas, these authors underline the primarily ethical nature of justification, as in these assertions: "the ethical relation renders ontology meaningful, not the other way around" (Child et al, 1995, p.

183); and "In this conception [of knowledge] it is not truth claims but ethical claims that edify and open up possibilities of new actions and judgment (Masschelein, 1998, p. 611; see also Masschelein, 1998; and Säfström, 1999).

4. The question of relevance in an antifoundationalist philosophy of education

Obviously, those philosophers of education who reject the foundationalist model of justification no longer see philosophy of education's primary task as contributing to the formulation and justification of education's founding principles. This turning away from foundationalism seems particularly to affect the normative role of philosophy of education. For example, according to Henry Giroux, philosophy of education has lost "the possibility to speak for all of mankind" (1997, p. 151). Accordingly, a well-established conception of philosophy of education and its contribution to educational practice seems to have come under threat. This section describes how the philosophers of education I evaluated perceive the loss of relevance that is involved with antifoundationalism.

At the most general level, this loss of relevance derives from losing those instruments that can be used to justify any kind of prescriptive pretension - that is, once one rejects epistemologically privileged foundations and procedures. it is no longer possible to justify claims in such a way that any 'rational' being must accept them. Consequently, philosophy of education seems to end up in a situation in which it no longer holds the status of a "master discipline" (Peters & Lankshear, 1996, p. 3). Many philosophers point to the general loss of prescriptive potential that results from rejecting all forms of epistemological privilege (Blake et al 1998; and 2001; Fitzsimons & Smith, 2000; Greene, 1995; Hogan, 1998: Masschelein, 1998; Prior McCarthy, 1995; Säfström, 1999; Simpson. 2000).

Other authors provide more specific reasons for philosophy of education's loss of relevance. Some of them emphasize the loss of critical potential in particular (Biesta, 2001; Gur-Ze'ev, 1998; Lather, 1998; Masschelein, 1998; and 2000; Prior McCarthy, 1995; Weinstein, 1995). Biesta, for example, argues that the philosophy of education can no longer fulfill its critical task because the lack of foundations robs us of the fixed criteria evaluative argumentation seems to require (2001, p. 136). Heyting also relates the problem of determining criteria for "finally settling an issue" to the problem of finding epistemologically privileged foundations, concluding that it hardly seems possible to formulate compelling criteria for critique (2001, p. 110). Even alternative solutions, such as making use of utopias for founding critique, seem unworkable, because these alternatives are considered as liable to power-knowledge or ethics-knowledge connections as any other kind of criteria (cf. Gur-Ze'ev, 1998; Masschelein, 1998).

Some authors are even more specific, interpreting the general loss of prescriptive and critical potential primarily in terms of social justice. Here again, the interrelatedness of epistemological and social-philosophical approaches appears to be important - emphasis on this point is a dominant characteristic of the texts I analyzed. As in the previously cited argument by Blake et al., for example, Biesta (1998) in particular observes a loss of potential to challenge power positions. Because he considers knowledge intrinsically related to power, he concludes, "Knowledge can no longer be used to combat power. "It...signifies the end of the 'innocence' of knowledge as a critical instrument" (ibid., p. 506)[4]. Kohli also interprets the loss of relevance from a perspective of social justice, but she understands social justice in terms of emancipation (Kohli, 1998). In her view, rationality can no longer guarantee a trouble-free critical procedure because it has lost its independent character. Weinstein approaches social justice in terms of a knowledge-perspective that takes diversity into consideration and that is committed to preventing exclusion. He shares Kohli's conclusion that rationality - robbed of its alleged independence - is not the right instrument for realizing this goal (Weinstein, 1995).

5. Contextual justification

The vast majority of the philosophers of education evaluated here highlight the social or personal embedding of knowledge and justification - a point that is important to understanding their general view of justification. In fact, the alternative views of justification that these authors recommend as appropriate replacements for the foundationalist model share the following characteristic: they emphasize the inter-relatedness of justifying knowledge claims and the contexts in which this process takes place. The texts I analyzed repeatedly present justification as a context-dependent process. In their alternative conceptions, these authors still treat justification as a process of giving reasons; however, they reject the hierarchical foundationalist model of justification that presupposes 'last' grounds and procedures that support all other grounds and procedures. Rather, they maintain that the validity of reasons, and of the processes in which giving reason takes shape, will vary based on the context. These alternative conceptions of justification all do without a firm

---

4   See also Blake et al (1998); they suggest that we may find ways of using
     knowledge to combat power, but not by securing foundations.

ground beyond all contexts-that is, a transcending meta-context. In this section, I will focus on the kinds of contexts these antifoundationalists take to be relevant and in what way. After that, I will look more closely at the 'contextualist model of justification' that emerges from these texts.

5.1 Context as meaning-context

One group of scholars appears to conceive of context as a contingent system of interrelated signs, on which any meaning is considered dependent. The context-dependency of justification naturally follows from this line of argument. These authors frequently refer to the philosophy of Jacques Derrida, specifically his notion of 'difference' (Biesta, 1998; and, 2001; Egea-Kuehne, 1995; Greene, 1995; Lather, 1998; Parker, 1997). On this view, meaning is not based in external reality; rather, the process of drawing distinctions or differences is the decisive factor in attaching meaning to objects or events (cf. Biesta, 2001). Consequently, the concept of 'justification' itself can only be understood as the result of a contingent distinction. Peters and Marshall (1999) suggest a similar approach (see also Peters, 1995). Following Lyotard's radical interpretation of Wittgenstein's later work, and with special emphasis on Lyotard's notion of 'le différend', they situate contexts of meaning in language games. 'Le différend' refers to the heterogeneity of the 'discursive universe', which cannot be resolved due to the lack of transcendent arguments or rules (Caroll, 1998). By making meaning dependent on coincidental and unstable meaning-contexts, Peters and Marshall emphasize the contingency of meaning and, as a consequence, restrict the validity of justification to a specific linguistic context.

5.2 Context as personal context

The second way of contextualizing justification conceives of it as a process deeply embedded in personal viewpoints. In this approach, the reasons that can be used to justify assertions should be understood as related to the personal point of view of the speaker. I found three versions of this perspective. The first one - attributable to the philosophy of Levinas - emphasizes the personal 'knowledge-horizon' (Child et al, 1995; Masschelein, 1998; and 2000; Säfström, 1999). In this perspective, the 'totality of things' (including justifying reasons) can only take shape within the horizon of a knowing person. Stated differently, this personal horizon constitutes the context for establishing justification, and, as Säfström argues, "It follows from this that speech about the other - the unfamiliar, conceptualized in a language which emanates from the knower in the center of the world-by necessity is caught up in a reduction of the other to the same" (1999, p. 227). Consequently, the significance of justification is considered restricted to a personal horizon.

In a second version, personal commitment, rather than the personal knowledge-horizon, is considered to be the relevant context for understanding justification. As a consequence, every justification is politically charged and in that sense restricted (Fitzsimons & Smith, 2000; Giroux, 1997; Peters & Lankshear, 1996; Prior McCarthy, 1995; Stone, 1995; Weinstein, 1995). In a fairly radical statement of this position, McLaren and Giroux maintain that "every time we use language, we engage in a highly partisan sociopolitical act...because each time we use it, we embody how cultural processes have been written on us and how we in turn write and produce our own scripts for naming and negotiating reality" (1997, p. 23). The third perspective that relates justification to personal context is sometimes called 'feminist'. Philosophers who adopt this stance relate justification to the personal corporality, rather than to a personal horizon or commitment. They consider processes of knowing and justification to be embedded in the 'lived body', which has its own historical and social setting. Prior McCarthy expresses this viewpoint in asserting that all knowledge is eventually 'constituted' in what she calls "features of our bodily selves" Prior McCarthy, 1995, p. 35). In a similar context, Kohli speaks of the "embodied selves" (1998, p. 519). According to this view, our corporal embedding restricts the significance of justification.

5.3 Context as discourse-context
The third and last type of context to which the antifoundationalists I considered relate justification consists in a social and communicative embedding (Blake et al, 1998; Heyting, 2001; Ruhloff, 2001; Simpson, 2000). According to this view, reasons and their relevance are related to the communicative discourse-context. Blake et al. define discourse as "a collection of statements [involving knowledge or validity claims] generated at a variety of times and places...and which hangs together according to certain principles as a unitary collection of statements" (1998, p. 14). According to this definition, the validity of grounds of justification is related to - and restricted by - the discourse in which they are (or can be) employed; consequently, any actual justification presup-poses a specific discourse in which this justification can be considered convincing (cf. Heyting, 2001). As opposed to the founda-tionalist model of justification, the authors who take this view do not conceive of a coordinating discourse-context that transcends the limitations of separate contexts with respect to justification.

To sum up, a theme that unifies the antifoundationalist materials I reviewed is their emphasis on the restricted validity of any justification. The way in which they relate justification to (different types of) context distinguishes this approach from (weak) foundationalism, on the one hand, and relativism, on the other. Although weak foundationalism does not require absolute certainty for basic foundations in the justificatory hierarchy, it holds on to the idea of relatively undisputed basic convictions that can serve justificatory ends irrespective of contextual factors. On this view, justification depends exclusively on content (compare with Williams, 2001, p. 164 ff.). By contrast, my sample of antifoundationalist philosophers rejects this idea of convictions endorsed independent of context and put into action whenever needed. As we have seen, these authors consider the validity of all ideas to be 'embedded'. or contextualized. Accordingly, justification – in terms of content as well as procedure – depends on contextually varying sets of assumptions that can be called into question in any other context.

This radical contextualist approach does not necessarily make these authors relativists either, however, because they do not think the context (whether semantical, personal, or discursive) makes things true[5]. According to these critics of foundationalism, justification is not passively subject to a preceding context that determines what propositions are to be considered 'privileged'. Although relativism relates epistemological privilege to a context, it continues to rely on the foundationalist model of justification. in which justification consists in hierarchically reducing claims to (now context-dependent) privileged basic convictions. Antifoundationalists also consider justification a process of reducing claims to undisputed ones, but unlike relativists, they do not consider the undisputed claims to be (part of) frameworks that are themselves beyond justification. Rather, they consider such claims to be ones that are momentarily taken for granted, but that can be brought up for discussion at any time (cf. Williams, 2001, pp. 226ff.). The preceding quote from McLaren and Giroux illustrates this point clearly by stressing that using language not only expresses the embodiment of cultural processes, but at the same time implies writing and producing our own scripts. Here, justification is de-

---

5   I borrow the term 'radical contextualist' from Williams (2001). For a basic definition of relativism, see Putnam's *Renewing philosophy*: "Truth in a language - any language - is determined by what the majority of the speakers of that language would say" (1992, P. 67). In his Problems of knowledge, Williams adds other possible frames to this definition by speaking of "relativism, which says that things are only 'true for' a particular person or 'culture' (2001, p. 10).

picted as active engagement in a specific context, thus indicating that being embedded does not imply being determined.

This contextualist approach affects the meaning and significance of 'justification' in a fundamental way: it stresses the local nature and relevance of any justification, thus robbing it of its traditional status and importance but not of its local relevance. This view emphasizes the aporetic nature of justification - that is, being justified no longer implies an unqualified recommendation but, rather, draws attention to the restricted nature of any recommendation. This point becomes clear when we examine what these authors think about the practical relevance of an antifoundationalist philosophy of education.

6. Benefits for the relevance of the philosophy of education

Although the antifoundationalist critics I evaluated for this study pay much attention to the relevance of philosophy of education, they do not seem to consider justification a decisive factor in the contribution to educational practice. These authors do not approach the question of philosophy of education's relevance from the perspective of epistemological considerations; instead, they reason primarily from social-philosophical considerations. As noted in the preceding discussion, these authors observed that rejecting the foundationalist model of justification results in a loss of prescriptive and critical potential; however, they do not appear to regret this 'loss'. Instead of trying to compensate for the limited scope of justification, antifoundationalists intend to turn this limitation into an advantage. They do not envisage another form of universalism or general validity, but instead take the lack of these options as their point of departure. The views I observed regarding the benefits of antifoundationalist approaches in philosophy of education appear to be related to the three justification contexts I distinguished in the previous section. According to these authors, philosophy of education's relevance results from making the restrictions that are inherent in each of those conceptions of context productive in a social-political sense.

6.1 Drawing attention to alternative meanings

Those antifoundationalists who consider justification in the light of contingent contexts of meaning emphasize the inevitable 'blind spot' of every context. In their view relevance should consist in preventing possible alternative meanings from being excluded as a consequence of such blind spots. Given that each meaning-context favors a restricted range of meanings, it hides all possible alternative meanings from view. Since these authors do not allow for compelling criteria for choosing the 'best' (or even a 'better') meaning-context in any situation, choosing a

41

context becomes a paramount responsibility. Antifoundationalists who associate themselves with concepts such as Derrida's 'deconstruction' (Biesta, 1998; and 2001; Egea-Kuehne, 1995; Greene, 1995; Lather, 1998; Parker, 1997), or Lyotard's 'legitimacy by paralogy' (Peters, 1995), share this approach. They contend that philosophy is in no position to take over this responsibility for choosing a meaning-context from educators. Ilan Gur-Ze'ev calls this philosophical inability to decisively make - and thus prescribe - such choices "philosophical negativism" (1998, p. 463). Instead of trying to release educators from the responsibility of having to choose the 'right' meaning-context, these authors reason the other way around, defining philosophy of education's relevance as its ability to enhance the awareness of this responsibility by pointing out the limited nature of any context and the resulting problem of choice. As a consequence, relevance is not primarily related to taking unequivocal positions, but rather to 'indecisiveness'.

A similar reaction can be observed with respect to the previously mentioned loss of critical potential. Awareness that alternative positions might be excluded is given priority over advocating a particular view. As a consequence, any position one might take is put into perspective from the very beginning. After all, every meaning is susceptible to deconstruction (as only one side of the 'difference') and thus contingent (cf. Masschelein, 1998, p. 524). The prior relevance of philosophy of education consists in bringing to the surface any meanings one inclines to take for granted (cf. Biesta, 2001; Greene, 1995; Gur- ze'ev, 1998; Parker, 1997) - a process that creates space for diversity, for the 'other' (cf. Biesta, 1998; Blake et al, 2001; Peters, 1995; Peters & Marshall, 1999).

6.2 Attention for the other person

Philosophers who relate justification to personal context represent a second perspective on the relevance of an antifoundationalist philosophy of education. They are also intent on resisting exclusion, not primarily in terms of 'meaning' but rather in a social-political sense. On this view, the concept of 'exclusion' is related to personal ways of life that are in danger of being excluded if specific contexts of justification are made too absolute and the relevance of personal contexts is neglected. I found two versions of this approach: the first is typical of authors who emphasize Levinas's concept of a 'personal horizon', and the second is characteristic of those who interpret the personal context in terms of 'embodied subjects' (that is, the feminist group) or personal commitment. I will begin by discussing this second approach.

Assuming that societies tend to force specific - and thus exclusionary - points of view upon their participants, this group points out that preventing exclusion demands heightened awareness and continuous effort. Philosophy of education can contribute to preventing exclusion by criticizing those social structures that shape and constrain persons. The idea is not to develop a positive blueprint of society - after all, every model is necessarily 'exclusionary'. In fact, thinking about a 'just' society from this perspective can only be done in negative terms (that is, in terms of what one must avoid). Gur-Ze'ev's term "philosophical negativism" can also be employed in this respect. According to McLaren and Giroux, philosophy of education should be able to denounce social issues in the light of a 'better' (though unrealizable) future: "[T]he purpose of developing a critical language of schooling is not to describe the world more objectively, but to create a more ethically empowering world which encourages a greater awareness of the way in which power can be mobilized for the purposes of human liberation" (1997, p. 21).

Although all of the authors in my sample consider context essential to procedures of justification, this contextual constraint elicits criticism - in order to avoid exclusion - rather than acceptance. Unlike relativists, antifoundationalists do not treat context as an inviolable constraint. Instead, they seize on this constraint as an incitement to seek out the boundaries of context and, if possible, to shift them. Against this background, Padraig Hogan points out the importance of dialogue: "It is crucial to realize that we need not just be prisoners, or helpless victims, of ... partiality" (1998, p. 370). Other authors attach a similar function to counternarratives (stories from marginalized points of view), which can contribute to overcoming the limitations of a particular context (Giroux, 1997; Peters & Lankshear, 1996). This approach sees philosophy of education as relevant primarily through its capacity to draw attention to these possibilities by finding and exploring examples of such counternarratives.

To this point, I have observed antifoundationalists' strong inclination to shift priority from epistemological questions toward social-philosophical ones. This tendency seems even more pronounced in those who, following Levinas, define personal context as 'personal horizon". Instead of the traditional objectifying conception of knowledge (in which we 'appropriate' the world as it is), these authors emphasize the constitutive and restrictive role of the personal horizon. In order to overcome the detrimental consequences of this restriction, they advocate a conception of knowledge as originating from "the ethical relation between self and other [person]" (Child et al., 1995, p. 183). According to them, one should let go of the restrictions that spring from such concepts as 'autonomy' and 'autonomous knowledge' and surrender to the heteronomy of the encounter with the other person. These authors argue in favor of openness to what 'exists' outside the personal horizon and remains unimaginable from any

43

'autonomous' position. Their ideas regarding how to escape the limitations of a personal perspective highlight the extent to which ethical or social-philosophical considerations have supplanted epistemological ones. They assert that such an escape is not effected by exchanging ideas or confronting meanings, but rather through the direct encounter with another human being. In the ethical experience central to such encounters, one is approached as an individual. This leaves the individual an individual-in-relation instead of a (potentially) autonomous one. In that capacity, the individual cannot escape from responsibility, from the "demand to judge and act upon the system in one's own name.... [H]uman beings are called upon as individuals, and only as individuals, to resist hegemony" (Masschelein, 1998. p. 529). In this approach, the relevance of philosophy of education consists in resisting the hegemony of the personal horizon. When persons open themselves to the 'other' in an existential sense, they will be able to avoid having their judgments determined by their preliminary personal position.

6.3 Continuing the conversation: questioning the boundaries of discourse-contexts

Finally, I turn to the relevance of a philosophy of education that interprets the context of justification as a discourse-context. Like those authors who see justification in terms of personal context, these critics also emphasize restrictive and hence exclusionary influences exerted by specific justification contexts (in this case, discourse-contexts); they expect to benefit by making people aware of these restrictions without intending to formulate a normative framework for standards of justification. Assuming that one is not necessarily imprisoned within the boundaries of actual discourse, these authors see philosophy of education's primary task as making explicit and calling into question those conventions that people are inclined to take for granted and that result in exclusionary practices. This makes it possible to challenge the constraints of discourse-contexts, push them and shift them. In this way, separate discourses can mutually enrich one another. Because this group of antifoundationalists does not pursue a fixed objective, the image that results is that of a philosophy of education that actively contributes to "continuing the conversation" about educational issues (Rorty, 1979).

7. A new normativity in philosophy of education

In analyzing the results of my inquiry, I observed that the contextual nature of justification became the 'leitmotiv'. The antifoundationalist approaches to philosophy of education I found in my sample appear to share the intention to avoid any claim to universal validity, not only for epistemological reasons, but also - perhaps even primarily - because of the exclusions that result from such claims. This raises two questions: First, is it possible to avoid any appeal to universality? And, second, even if we determine that it is possible to avoid such an appeal, how should we judge the relevance of this kind of philosophy of education?

I will address the first question with reference to two critical arguments made by Siegel. He maintains, first, that it is not possible to avoid universal claims with respect to basic points of departure and, second, that it is not possible to avoid universal claims with respect to procedures of justification.

Child, Williams, and Birch provide a fairly representative characterization of how authors in my sample reject universal claims: "Claims of unity, certainty, and universality...are always taken up and protected by situated, embodied persons. They are thus...not possibly universal, ultimate, or certain" (Child et al. 1995, p. 167). According to Siegel, however, "in denying the universal one embraces it; one cannot escape the universal by denying it" - that is, denying universality is self-contradictory because it inevitably implies such a claim (1996, p. 174). The preceding quotation seems to confirm this, if only by using such terms as 'always' and 'not possible'. It is valid to ask whether such an argument necessarily makes my authors (weak) foundationalists after all.

Before answering this question, however, it is important to remember the primary characteristics of foundationalism as explained at the outset of this paper: a conception of knowledge as hierarchically structured that rests on a basis of relatively certain claims and/or is derived according to relatively privileged procedures. Thus understood, I doubt whether the arguments put forth in my sample can ultimately be reduced to building alternative hierarchies with the same characteristics. To start with, neither the shared repudiation of universal claims, nor the shared intention to avoid exclusion, functions as an undisputed ground for justifying other claims. These so-called 'propositional attitudes' do not owe their status to any epistemological features. In fact, the philosophers of education I studied point out that an important reason for resisting the primacy of epistemology concerns its practical consequences. As they see it, weak foundationalism is no option for resolving the problems that result from lacking absolute certainty (the traditional prerequisite for any foundationalist approach), because in the end such a solution amounts to uncontrolled acceptance of social exclusion. Instead of clinging to the primacy of epistemology by reformulating it (the approach characteristic of weak foundationalism), these authors develop epistemological views that deny the primacy of epistemology, putting a basic commitment - that is, to the different varieties of inclusion I described previously - in its

45

place. Compared to that commitment, justification is only of minor importance.

Given this degradation of epistemology in their conception of knowledge, antifoundationalist philosophers emphasize the local, contextual basis of any justification's acceptability. As a consequence, they do not judge or evaluate positions as such, but always as related to a context and with a view to their practical impact. Because assertions 'as such' lack any epistemological status, they contend, the process of justification cannot be performed if the context and the preliminary commitments related to this context are ignored. To return to Siegel, he contests this contextual restriction of validity claims, arguing that mathematics, for example, remains valid irrespective of whether its claims are accepted in a specific situation (ibid., p. 176). Siegel seems to consider relativism the only alternative for foundationalism; however, the anti-foundationalists I studied seem to view justification less as a matter of 'all or nothing' than foundationalist models assume. According to this view, denying universalism does not necessarily mean particularism, and denying objectivity does not necessarily imply subjectivism or social relativism. Rather, emphasizing the embedded nature of knowledge draws attention to the interactive dimensions of justification (see Burbules, 1995). To sum up, it may be said that the authors in my sample appeal to claims with a general import, but that does not make them (weak) foundationalists. In addition, their use of rational procedures (while denying their universal validity) does not make them foundationalists either. Although Siegel considers any appeal to rational procedures while denying their universal validity self-contradictory, such a verdict ignores the different meaning antifoundationalist philosophers attach to justificatory procedures (see Siegel, 1996, p. 180; and 1998, p. 30). In the first place, their validity is considered contextual in nature - as a consequence, philosophical or scientific justification does not necessarily imply practical justification. In the second place, in part because of their rejection of the primacy of epistemology, justification by means of rational procedures is considered of less importance. Kohli observes, for example, that our rationalist Cartesian tradition is tantamount to excluding "the emotional, the sensuous, the imaginative" (1998, p. 516).

Although it seems possible to avoid universal claims, and my authors cannot be labeled (weak) foundationalists, there remains the question of whether a philosophy of education that considers grounds only of the local, and justification only of limited importance, retains any relevance. Although the antifoundationalists' primary aim is to avoid exclusion, their approach does not allow for generalizing this aim either. On closer inspection, such generalization is not what they intend. Rather than plead in favor of a specific position, they want to create openness to a plurality of possible positions. However, this intention does not negate the fact that their position radiates a strongly normative appeal, that favors pluralism. In the absence of epistemologically justified foundations and procedures, the philosopher of education is ultimately thrown back on his or her commitment, which cannot be founded but which still emanates normativity. In conclusion, the characteristics of this normativity deserve closer examination.

In a way, the personally or discursively embedded commitment emphasized in the texts I analyzed is reminiscent of the normative approach dominant in traditional continental philosophy of education. This tradition, which prevailed through the 1950s, aimed at developing educational doctrines based on religious or political principles that were not considered subject to further justification, but instead were a matter of existential creed (cf. Stellwag, 1962). However, unlike this traditional normative philosophy of education, the kind of commitment emphasized by the antifoundationalist philosophers I studied does not serve the development of any doctrines, let alone prescriptive ones. For example, Lather writes that she aims not for approaches that will enable us to settle any educational dilemmas, but for approaches "that call out a promise of practice on a shifting ground" (1998, p. 497).

The antifoundationalist approach as I have reconstructed it also has similarities to a more recent kind of normativity in philosophy of education - that developed during the 1970s in a first wave of response to Habermas's critical theory. In this variety, procedures rather than first principles (whether religious or political) were of central importance. Furthermore, this trend in philosophy of education did not aim at developing any prescriptive doctrines; like contemporary critics analyzed here, critical philosophers of education of the 1970s aimed at stimulating (critical) attitudes and discourse. Unlike their contemporary counterparts, however, they did so with the intention of ensuring dominion-free consensus. Although power appears to be an important factor to a vast majority of my authors, they do not aim at defeating its influence, nor do they expect this to be possible. The publications analyzed here. then, seem to express a 'new' normativity that avoids both the formation of prescriptive doctrines and the formation of decisive procedures, as descrybed in the following: "This in turn involves building an ethos...where differences and tensions are not only acknowledged, but also experienced in such a way that they enlarge the context of what is taught and learned, enrich an appreciation of diversity and progressively discipline the preconceptions that underlie the exercise of all judgment" (Hogan, 1998. p. 371).

What is 'new' about the emerging normativity is its orientation toward openness and undecidability. According to my authors, philosophy of education should not put itself in the position of practitioners by taking over their responsibility for making decisions and choices. In their view, philosophers of education should stimulate and revive discussions rather than try to conclude them. From this perspective, human expression and inter-human exchange seem to be valued over any well-founded conclusion.

8. References

Biesta, G.J.J. (1998). Say you want a revolution...Suggestions for the impossible future of critical pedagogy, *Educational Theory, 48*(4), 499-510.

Biesta, G.J.J. (2001). How can philosophy of education be critical? How critical can philosophy of education be? Deconstructive reflections on children's rights, in: Frieda Heyting, Dieter Lenzen and John White (Eds.), *Methods in Philosophy of Education* (pp. 123-43). London: Routledge.

Blake, N., Smeyers, P., Smith, R. & Standish, P. (1998). *Thinking Again. Education after postmodernism*. London: Begin & Garvey.

Blake, N., Smeyers, P., Smith, R. & Standish, P. (2000). *Education in an age of nihilism*. London: Routledge.

Burbules, N. (1995). Reasonable doubt: toward a postmodern defence of reason as an educational aim, in: Wendy Kohli (Ed.), *Critical conversations in philosophy of education*. New York: Routledge.

Carrol, David (1998). Lyotard, Jean-Francois, *Routledge Encyclopedia of Philosophy*. London: Routledge.

Child, M., Williams, D.D. & Birch, J.A. (1995). Autonomy or heteronomy? Levinas's challenge to modernism and postmodernism, *Educational Theory, 45*(2), 167-89.

Copi, I. (1967). Language analysis and metaphysical inquiry, in: Richard Rorty (Ed.), *The linguistic turn* (pp. 127-32). Chicago: The University of Chicago Press.

Dancy, J. (1985). *An introduction to contemporary epistemology*. Oxford: Blackwell.

De Ruyter, D. (2003). The importance of ideals in education, *Journal of Philosophy of Education, 37*(3), 468-82.

Egea-Kuehne, D. (1995). Deconstruction revisited and Derrida's call for academic responsibility, *Educational Theory, 45*(3), 293-309.

Fitzsimons, P. & Smith, G. (2000). Philosophy and indigenous cultural transformation, *Educational Philosophy and Theory, 32*(1), 25-41.

Giroux, H. (1997). *Pedagogy and the politics of hope - Theory, culture, and schooling*. Boulder: Westview Press.

Gitlin, A. & Thompson, A. (1995). Creating space for reconstructing knowledge in feminist pedagogy, *Educational Theory, 45*(2), 125-50.

Greene, M. (1995). What counts as philosophy of education, in: Wendy Kohli (Ed.), *Critical conversation in philosophy of education* (pp. 3-23). New York: Routledge.

Gur-Ze'ev, I. (1998). Towards a non-repressive critical pedagogy, *Educational Theory, 48*(4), 463-86.

Heyting, F. (2001). Antifoundationalist foundational research - Analysing discourse in children's rights to decide, in: Frieda Heyting, Dieter Lenzen and John White (Eds.), *Methods in Philosophy of Education* (pp. 108-24). London: Routledge.

Hogan, P. (1998). Europe And the world of learning: Orthodoxy and aspiration in the wake of modernity, *Journal of Philosophy of Education, 32*(3), 361-76.

Kohli, W. (1998). Critical education and embodied subjects: Making the poststructural turn, *Educational Theory, 48*(4), 511-19.

Lather, P. (1998). Critical pedagogy and its complicities: A praxis of stuck places, *Educational Theory, 48*(4), 487-97.

Marshall, J.D. (1995). Wittgenstein and Foucault: Resolving philosophical puzzles, in: Paul Smeyers and James D. Marshall (Eds.), *Philosophy and education: Accepting Wittgenstein's challenge* (pp. 205-20). Dordrecht: Kluwer Academic Press.

Marshall, J.D. (1996). Michel Foucault: Personal autonomy and education, in: C.J.B. MacMillan & D.C. Philips (Eds.), *Philosophy and education (Vol. 7)* (pp. 511-19). Dordrecht: Kluwer Academic Press.

Masschelein, J. (1998). How to imagine something exterior to the system: Critical education as problematization, *Educational Theory, 48*(4), 521-30.

Masschelein, J. (2000). Can education still be critical? *Journal of Philosophy of Education, 34*(4), 603-16.

McLaren, P. & Giroux, H.A. (1997). Writing from the margins: Geographies of identity, pedagogy, and power, in: Peter McLaren (Ed.), *Revolutionary Multiculturalism - Pedagogies of dissent for the new millennium* (pp. 16-41). Boulder: Westview Press.

Parker, S. (1997). *Reflective teaching in the postmodern world: A manifesto for education in postmodernity*. Buckingham: Open University Press.

Peters, M. (1995). Education and the postmodern condition: Revisiting Jean-Francois Lyotard, *Journal of Philosophy of Education, 29*(3), 387-400.

Peters, M. & Lankshear C. (1996). Postmodern counternarratives, in: Henry A. Giroux, Peter McLaren & Michael Peters (Eds.), *Counternarratives - Cultural studies and critical pedagogies in postmodern spaces* (pp. 1-39). New York: Routledge.

Peters, M. & Marshall, J.D. (1999). *Wittgenstein: Philosophy, postmodernism, pedagogy.* Westport: Bergin & Garvey.

Prior McCarthy, L. (1995). Bodies of knowledge. *Studies in Philosophy and Education, 14,* 35-48

Putnam, H. (1992). *Renewing philosophy.* Cambridge, Mass.: Harvard University Press.

Rorty, R. (1979). *Philosophy and the mirror of nature.* Princeton: Princeton University Press.

Ruhloff, J. (2001). The problematic employment of reason in philosophy of Bildung and education, in: Frieda Heyting, Dieter Lenzen and John White (Eds.), *Methods in Philosophy of Education* (pp. 157-72). London: Routledge.

Säfström, C.A. (1999). On the way to a postmodern curriculum theory - moving from the question of unity to the question of difference, *Studies in Philosophy of Education, 18*(4), 221-33.

Siegel, H. (1987). Relativism refuted: A critique of contemporary epistemological relativism, in: *Synthese library: Studies in epistemology, logic, methodology, and philosophy of science.* Dordrecht: Reidel.

Siegel, H. (1996). *Rationality redeemed? Further dialogues on an educational ideal.* New York: Routledge.

Siegel, H. (1998). Knowledge, truth and education, in: David Carr (Ed.), *Education, knowledge and truth - Beyond the postmodern impasse.* New York: Routledge.

Simpson, E. (2000). Knowledge in the postmodern university, *Educational Theory, 50*(2), 157-77.

Smeyers, P. (2001). Taalfilosofie: Op weg naar een wijsgerige pedagogiek als een filosofie van een praxis, in: Paul Smeyers & Bas Levering (Eds.), *Grondslagen van de wetenschappelijke pedagogiek* (pp. 112-29). Amsterdam: Boom.

Snik, G. (1999). Grondslagen van liberale visies op onderwijsvrijheid, *Pedagogisch Tijdschrift, 24*(1), 125-51.

Snik, G., Haaften, W. van & Tellings, A. (1994). Pedagogisch grondslagenonderzoek, *Pedagogisch Tijdschrift, 19*(4), 287-304.

Stellwag, H.W.F. (1962). Positief. of negatief, *Pedagogische Studiën, 39*(4), 321-32.

Stone, L. (1995). Narrative in philosophy of education: A feminist tale of "uncertain" knowledge, in: Wendy Kohli (Ed.), *Critical conversations in philosophy of education* (pp.173-89). New York: Routledge.

Usher, R., Bryant, I. & Johnston, R. (1997). *Adult education and the postmodern challenge, learning beyond the limits*. London: Routledge.

Weinstein, M. (1995). Social justice, epistemology and educational reform. *Journal of Philosophy of Education, 29*(3), 369-85.

White, J. & White, P. (2001). An analytic perspective, in: Frieda Heyting, Dieter Lenzen and John White (Eds.), *Methods in Philosophy of Education* (pp. 13-29). London: Routledge.

Williams, M. (2001). *Problems of knowledge: a critical introduction to epistemology*. Oxford: Oxford University Press.

Wittgenstein, L. (2001). *Philosophical investigations* (G.E.M. Anscombe, Trans.). Oxford: Blackwell.

III
EPISTEMOLOGICAL INSIGHTS AND CONSEQUENCES FOR PHILOSOPHY OF EDUCATION I: FOUNDATIONALISM, FALLIBILISM, AND CONTEXTUALISM

1. Introduction

To date, antifoundationalism plays a significant role in discussions about the future of philosophy of education. However, it certainly is not the only voice in the debate. It is indeed one of the most outspoken voices and does, perhaps the most radically, advocate the epistemic uncertainty that has affected philosophy of education over the past decades. On the other hand, it has been repudiated from the very beginning by a large group of, in an epistemological sense, more temperate philosophers of education. According to these authors, the current epistemological discussions do not at all have to lead to, what they regard as apocalyptic, ideas as proposed by antifoundationalists (cf. Carr, 1998a). This does, however, in no way imply that these authors are blind to epistemic uncertainty. It is just the way they interpret, and deal with, epistemic uncertainty that differs in principle from the antifoundationalists' position. It is important to also make the voice of this group of authors clearly heard when it comes to my search for a better understanding of the epistemic uncertainty we seem to be confronted with. In this chapter, I will work my way towards doing so by first briefly touching upon the central question of epistemology, and why it is of importance to education. I will subsequently go into the background of the epistemic uncertainty with which we are faced, as a basis for showing how this uncertainty is dealt with by a vast amount of non-antifoundationalist philosophers of education. This should eventually lead to an assessment of both the fruitful and more problematic issues regarding the different ways of dealing with epistemic uncertainty that will have passed in review, based on which I will be able to continue my search for my own position in the debate.

Epistemology revolves around knowledge. It is especially about the question of what 'knowledge' is, and when one may speak of it. For various reasons, these questions have always played a key role in Western philosophy (cf. Elgin 1996, p. 22). The possession of knowledge has been regarded as valuable to individual life ever since antiquity. A life directed by knowledge reflects an ideal of self-control, as opposed to a life that runs as a matter-of-course in line with traditions and customs or ill-considered, everyday ideas. Socrates' statement that an unexplored life is not worth living can also be understood in these terms. There are also other considerations, which are more related to the role that knowledge plays in the public domain. Knowledge is deemed to fulfil an important instrumental value, for instance. Efficient actions based on knowledge are attributed a higher success rate than actions that are based on one's own discretion or trial and error. Partly for these reasons, 'having' knowledge is also considered an honorary title from which status and authority may be derived (cf. Williams 2001, p 11). In the public debate, someone who is considered to be knowledgeable will be heard sooner than someone who is regarded as ignorant, for example.

The attribution of knowledge thus generates status and clearly has practical implications. This makes the question of who possesses knowledge directly relevant to a subject area like education. We expect teachers to transfer knowledge to their pupils, for example. From professional educational practitioners, it is often demanded that their teaching practices are in line with current scientific knowledge of education. Lastly, chapter one reveals that, since ancient times, it has also been deemed important to ascertain to what extent philosophers of education can lay claim to making statements about knowledge, because these had consequences for the potential practical relevance of the philosophy of education. The fact that the possession, or lack, of knowledge has direct practical implications also makes the attribution of that knowledge a normative affair that should be handled carefully. Pivotal is the question under which circumstances we tend to recognize someone's statements as statements of knowledge. It is primarily this epistemological issue that will be central to the rest of this chapter. In other words, I will focus on the question of when we deem someone justified to lay claim to knowledge or – to use Williams' terminology – when we may refer to 'epistemic entitlement' (Williams 2001, p. 21).

2. The fruitless search for infallible knowledge

According to classic foundationalism[6], which for a long time has been the dominant epistemological position within Western philosophy, 'epistemic entitlement' can only be spoken of when basically there is no chance that the claim to knowledge at hand is unjustified (cf. Elgin 1999, p. 22). Here, unjustified means that the claim is not in accordance with external reality. In such cases, knowledge is by definition considered as infallible and universally valid. In itself, the ambition that classic foundationalism sets itself is very noble, particularly in light of the previously mentioned normative effect of the attribution of knowledge. Nowadays, however, there will scarcely be a philosopher who thinks this to be a feasible ambition, since hardly anyone will believe that the requirements set within the

---

6   I borrow this term from Jonathan Dancy (cf. 1985, p. 53).

epistemological approach to the justification of claims, can be met. I will first look into these requirements.

The question of 'epistemic entitlement' is always dependent on the degree to which a certain belief seems acceptable, or justified, in the light of something else. To a foundationalist, the proper justification of a belief is only the case if it is supported by a belief or authority that itself is also justified. It also applies to foundationalism that statements that are used for justification have to be epistemologically more certain than the statements they support. As such, they should be at a deeper, more fundamental level. To foundationalists, knowledge justification necessarily implies a hierarchical order of levels of beliefs, where deeper layers should be gradually more certain than those above. This vertical order of the justification chain makes foundationalists susceptible to the pitfall of infinite regression. After all, each statement is for its own justification dependent on a deeper lying statement, which in turn needs justification on an even deeper level. In order to avoid this pitfall, the justification chain eventually needs a stopper. According to classic foundationalism, a proper justification chain can only be brought to a halt by beliefs or authorities that can serve as support for other statements, but that are not dependent on other beliefs for their own justification - the so-called basic beliefs, or foundations. The characteristic of these foundations, therefore, is that they are justified *in themselves*. Since these foundations need no further justification, they are in epistemological perspective one up. I have typified them in that sense as 'epistemologically privileged' in chapter two.

Precisely regarding the issue of presupposing justified foundations that themselves need no further justification, classic foundationalism has been heavily criticized. Several philosophers have pointed out that the most important mental events that have traditionally been designated as possible candidates for supplying an epistemological foundation, to wit human reason, the sensory experience, and language, have been found to be unsuitable (cf. Baynes 1996; Rorty 1989). This criticism of classic foundationalism is not voiced by merely a small faction in philosophy, but is now widely shared. In chapter two, it became clear that the same criticism can also be heard among philosophers of education. Within the philosophy of education it is not just a small group of more postmodern inclined authors either, who have departed from strict classic foundationalism, like the antifoundationalists in chapter two. The notion that there are no authorities to be found that we can rely upon to arrive at infallible, universally valid knowledge, is in fact shared in contemporary philosophy of education (see also Adler 2003; Blake et al. 1998; Carr 1998a).

Thus, at second glance, the essential candidates for supplying the foundations for infallible knowledge do not seem to be equipped for the task. Against the background of the classic foundationalist idea that we can only speak of a justified claim to knowledge when there is no possibility of that claim being unjustified, it seems that we can only draw the conclusion that it is beyond human capacity to ever arrive at 'real' knowledge. In other words, if we have to let go of the notion of foundations that are justified in themselves, we also have to let go of the idea of eternally guaranteed, accurate - or infallible - knowledge. This makes it clear as to why epistemology will have to reconcile with the notion that the justification of claims inevitably entails a certain degree of uncertainty. The subsequent questions are: how to interpret that uncertainty and how to deal with it epistemologically.

3. General validity without infallibility

To many philosophers, this epistemic uncertainty does not at all imply that we may no longer refer to 'real' knowledge, or that we are no longer able to defend certain beliefs over others. On the contrary, a great number of philosophers share the conviction that, even without epistemological foundations that are justified in themselves, it is still possible to arrive at objective knowledge. The idea is that, even though no single belief may be infallible, in certain cases we may have sufficient reasons to classify some beliefs as in any case the best possible of those currently available to us and, therefore, may regard them – at least for now – as ultimate, generally valid basic beliefs that can be used to ground our knowledge claims. Due to the recognition of the fallibility of such an epistemological ground, this position is also referred to as fallibilist. Within this approach we have to take into account that the basic beliefs we rely upon for justification may over time become outdated. Against this background, fallibilist authors stress that we will constantly have to be critical of the beliefs we adhere to, so as to check their current tenability. This so-called fallibilism is something we often come across within the philosophy of education. In order to examine what this position would actually entail, I am presenting two educational philosophical examples that are relevant: 1) because it concerns influential philosophers of education; and 2) because they focus on different domains of the judgment; and 3) because their interpretation of fallibilism differs.

The first example can be found in the work by Harvey Siegel, who extensively discussed the relationship between epistemology and (philosophy of) education (cf. Siegel 1990; 1997; 1998). To Siegel, the ultimate ground for epistemological justification of knowledge is the absolute validity of rationality. This basic belief is, in his view, inevitable, since rationality is presupposed in the practice of inter-human communication. According to Siegel, by bringing the validity of rationality up for discussion we unavoidably end up in a pragmatic contradiction, because we appeal to rational considerations by bringing it up for discussion in the first place (cf. Siegel 1996, pp. 81-82). Despite his claim to absolute validity of rationality, Siegel may be regarded as a fallibilist, for in his eyes we should recognize that the actual criteria that we use for the rational evaluation of judgments, are fallible (ibid, pp. 176-178). These criteria are actually the formal and informal logical standards that we apply in, what Siegel refers to as, critical thinking; whether, for instance, we base our arguments as much as possible on relevant and correct information, or whether we do not draw conclusions that cannot be directly deduced from premises (cf. ibid., p. 95). Although such standards are deemed fallible, according to Siegel we may suppose that they are the best possible available to us at this moment, so that an objective validity may also be attributed to them. We may do so, because these criteria themselves are also continually subject to critical assessment. After all, according to Siegel, critical thinking obliges us to critically assess not only our own judgments, but also the standards that we use for that critical assessment, precisely because of our recognition of their fallibility (cf. ibid.). This makes it clear as to how we, according to Siegel, can safe-guard the general validity of our epistemological basic beliefs even if we assume their potential fallibility, whereby the objectivity of knowledge claims would also be safeguarded.

The second example is specifically focused on the justification of moral judgments and can be found in the work by David Carr on the objectivity of values (cf. Carr 1998b; Carr 2000). According to Carr, moral judgment always revolves around the improvement of human well-being, avoiding sorrow, or battling evil. Carr submits that we must acknowledge that the way in which we interpret this at this moment, will probably not prove to be correct once and for all. In his view, our moral judgments are fallible in this regard. This, Carr stresses, does not at all mean that we would not be able to arrive at moral objectivity or even the truth and that we should acknowledge, for example, that morality is relative to the community in which we happen to live – which he believes is argued by postmodern communitarianist oriented philosophers. He opines that we are most certainly able, for instance, to objectively determine that issues, like slavery, the burning of women, and child murder are morally reprehensible. We can so determine, Carr continues, on the basis of moral investigation, in which we focus, among other things, on concrete, perceptible facts that concretely show which well-being or sorrow is induced by certain actions (ibid., p. 24). With this very statement Carr also opposes modern liberalist philosophers who believe that moral objectivity might be realized on the basis of the strict, rational inference of prescriptive, moral basic beliefs. In this light, we will morally condemn slavery, not so much because it goes against a rationally inferred principle of human self-determination, but rather because we have simply been able to determine which sorrow it has caused, both on an individual and collective level. Partly due to matters like scientific progress and open (e.g. cross-cultural) communication, it would be possible to gain increasingly nuanced insight into the reality of what is morally correct, or incorrect, in the way we practice. According to Carr, this also implies that in the course of time we will have to adapt our current moral judgments. "[J]ust as we know – in the light of what human progress there has been – that past personal and social moral sensibilities seriously failed to register real human needs and interests, so we can be sure that other moral needs, currently beyond our present ken, await discernment through further sensitive reflection and interpersonal engagement" (ibid., p. 125). However, for the moment we do not doubt our basic moral beliefs. We do know, Carr continues, which cases are morally valuable. If we continue to carefully examine whether these values are met, we will also realize when it is necessary to adapt our insights.

These examples have shown how fallibilism, which many philosophers of education regard as a acceptable and constructive way of dealing with the inevitable epistemic uncertainty we face, may be interpreted. We should take some time to further explore fallibilism as an approach in order to also determine its potential value for the philosophy of education. First of all, I would like to address two clearly powerful aspects of fallibilism.

The first strength of fallibilism lies in the fact that it urges to constantly look for arguments or examples that might lead to adapt, or even replace, current basic epistemological grounds in order to raise the level of judgment increasingly higher. The most renown example of this practice may be found in the Popperian scientist who realizes that his theory is a provisional theory and accepts the responsibility to continue to search for contradicting observations, so as to eventually be able to contribute to the expansion and refinement of the body of knowledge. This self-searching and self-correcting intention also shows that fallibilists - as do classic foundationalists - are trying very hard to forestall the attribution of unjustified claims. 'Epistemic entitlement' only applies once a claim is supported by grounds for justification for which we have sufficient reasons to state that they are correct, or that they are at least the best ones available. A criticaster could now argue that this vigilance does not alter the fact that the fallibilist's principles are uncertain and that, therefore, we still cannot know for sure whether our knowledge attributions are indeed correct. A fallibilist will probably not lose any sleep over this criticism.

They would assert that there are simply no more definite principles, and that we all appeal to some basic beliefs sooner or later. Against that background, it seems most reasonable to at least acknowledge the uncertainty of those principles, before reviewing them critically and continually.

I believe the second strength of fallibilism concerns the possibilities it offers for positive theoretical judgment or, to put it in other words, the capability to take a firm stance. Fallibilists differ in principle on this specific subject from the antifoundationalists in chapter two who suggest that philosophers of education's primary concern should not be the defending of a certain position, since every defense is bound by restrictions and, as such, would have an excluding effect. The notion that we have access to basic beliefs that are reliable enough to be attributed a general validity to, immediately generates the possibility to more or less objectively defend, or indeed discard, certain insights. This is not only relevant in an epistemological sense, but also has consequences for the practical relevance of the philosophy of education. This can also be found in both Siegel's and Carr's fallibilism. Siegel asserts that based on the application of rationality criteria, we may for instance objectively defend critical thinking as an educational goal (1996, pp. 23-25), or expose our prejudices as irrational (ibid., p. 95). In this light, Siegel believes that the rationality criteria "also function as criteria for the evaluation of social organization and of procedures for the establishment of social and public policy" (ibid., p. 97). Carr shows us how, based on the idea of objectivity of values, a specific interpretation of moral education can be defended, or how it helps us to morally discard, or actually defend, specific practices (cf. Carr 1998b; 2000). As such, fallibilism fits in with a long-standing tradition in which educational philosophical insights are regarded as directives for educational practice (see chapter one).

Thus, fallibilism reveals itself to be a fruitful, and in that sense defendable, epistemological position for philosophy of education. However, if we zoom in on fallibilism as an epistemological stance, the approach raises some questions relevant to my inquiry into the possible merits of an acceptable contemporary epistemological approach. These questions concern, among other things, the relationship between the beliefs that justification is grounded in and the beliefs that need justification, as proposed by fallibilists. In fallibilism, justificatory grounds derive their status from the assumption that they are more reliable, or more certain, than the beliefs they are to support. We can thus make a distinction between more certain, justifying beliefs and less certain beliefs that need justification. Therefore, fallibilism has a vertical, or hierarchical, justification structure. We have already seen that such a justification structure is susceptible to the pitfall of infinite regression. To prevent this, the chain of justification needs a stopper. In fallibilism, this stopper is supplied by beliefs that are regarded as more certain and have been designated as epistemologically basic. As such, the epistemological principles of fallibilism fulfill the same role as the foundations of classic foundationalism, albeit that the requirements set for whatever should eventually serve as a foundation, are not as strict. That is why fallibilism is also described as mild, or lenient, foundationalism (Elgin 1996), or as foundationalism without infallibility (Dancy 1989). Be it as it may, both epistemological approaches have in common that they are based on ultimate, deeper lying justificatory grounds, that - for now, at least - need no further justification, hence the reason for regarding them as 'epistemologically privileged'. The advantage of fallibilism over classic foundationalism lies in the fact that it does not need to appeal to the notion of foundations that are justified in themselves, which emerged as unattainable. Still, even this notion raises questions.

Firstly, we may ask ourselves what the epistemological value may be of beliefs that are underpinned by ultimate basic beliefs that we recognize as fallible. Beliefs that function as a stopper in the justification chain are not under debate - at least not at the moment. Fallibilists argue that these more basic beliefs have been critically reviewed at other times. We thus have well-founded reasons to accept them as such. However, this does not take away the fact that they are now regarded as 'true', while at the same time recognized as fallible. This means that such epistemological basic beliefs actually appear as prejudiced. This does not necessarily have to be a problem. After all, we must accept that all our beliefs are fallible. Hence, this also applies to the basic beliefs that we use to justify our claims. The point here, however, is that the idea of fallible basic beliefs simply does not go together with the fallibilist pretence that we can still arrive at generally valid, or even objective, claims to knowledge. This may be clarified by looking into the beliefs that were once earmarked as epistemological principles, but lost that qualification later, when better alternatives became available. An example: we may readily accept that Ptolemy and his followers had good reasons to belief that the earth is the centre of the universe as their starting point in their study of celestial bodies. With this idea as the point of departure, various important discoveries were made, making it extremely influential from a scientific point of view. However, in light of later understandings, such as Copernicus' heliocentric view of the universe, or Einstein's theory of general relativity, it seems absurd to say that people in those days would have been justified – in epistemological sense – to claim general validity of the idea (for instance, it would imply that Galileo's claims, in an epistemological sense, were less correct). On second thought, the idea was really not as reliable or even as certain as people once believed, so that in retrospect we can say that a claim to general validity would have been unjustified.

In connection with moral principles, which Carr focuses on, we may be able to understand, considering the circumstances and insights at the time, why the ancient Greek defended the idea that certain members of society – such as slaves or women – were simply less valuable than others. However, it does seem strange to accept that they would have been allowed to claim the general validity of that claim in those days. However, it is quite plausible that they had no doubts about the idea whatsoever when it came to the arrangement of their society, and thus regarded it as the best possible idea available to them. What it boils down to is, if we now recognise that our principles are fallible, we also recognise they could be different (and most likely *will* change again in the future). This implies that we can hardly defend that anyone at this moment should reasonably embrace these principles as generally valid, even though there are quite good reasons to take them for granted, and even though we cannot imagine now that there could be a better alternative. The problem, therefore, does not necessarily lie in the defensibility of the justification model that is characteristic to fallibilism, but primarily in the epistemological status connected with this model. If you want to embed justification in ultimate beliefs whilst recognising their fallibility, then it seems inevitable that the pretence of general validity should be abandoned - if only because we will always have to take into account that there might be a better candidate that could serve as ultimate justificatory ground.

The notion of alternative candidates that could possibly serve as epistemological principles raises another question: How can we actually establish the status of a certain belief as the best possible alternative for serving as the ultimate ground for the justification of knowledge? To put it differently, how do we actually come to the attribution of epistemological privilege? This does not appear to be an easy task. If epistemologically privileged beliefs are the ultimate stopper in the chain of justification, how do we infer the criterion for the attribution of the privilege? If we regard the judgment that needs to be passed in order to clinch that attribution as a knowledge statement, this judgment must at least be supported in an epistemological sense. In line with fallibilist reasoning, this requires that we appeal to the best possible epistemological principles available to us at this moment. This would in fact mean that we evaluate the tenability of our current basic beliefs in light of those very same beliefs. Let us again turn to Siegel, who uses critical thinking to scrutinise the criteria for critical thinking itself in order to make sure it concerns the best possible grounds for the justification of knowledge. Whatever the case, it does seem impossible here to apply all the criteria for critical thinking in the evaluation of critical thinking as a whole. It may remind us of the well-known metaphor of dragging yourself out of the swamp by pulling your own hair. We must accept that a critical evaluation of critical thinking based on that very same critical thinking will never be able to lead to the discarding of critical thinking as a whole. After all, in such an evaluation we are continually forced to call upon our critical thinking itself, making it a presupposed epistemological principle. We could, however, imagine that we evaluate certain aspects of critical thinking on the basis of other aspects of critical thinking – holding certain rationality criteria in the light of other rationality criteria, for instance. Those aspects should in turn be evaluated on the basis of others, et cetera. Such an evaluation may lead to an adaptation, or honing, of the logical-argumentative criteria applied within critical thinking. However, it again applies that this form of justification rests on fallible principles, so that no general validity may be ascribed to these conclusions, even if we have good reasons to use them to justify our beliefs. This line of argumentation shows that it is difficult to understand how we could ever arrive at a replacement of our basic epistemological beliefs, simply because we will have to appeal to the basic beliefs that we currently apply in order to do so. On the other hand, we did see that this does not mean that we cannot arrive at an adaptation of beliefs. Epistemologically, however, such an adaptation or honing cannot lead to a claim of general validity, since the adaptation is grounded in fallible beliefs.

Fallibilism, so it shows, offers us a model for the justification of beliefs that allows room for epistemic uncertainty. Without appealing to infallible foundations, fallibilism offers us the possibility of defending certain claims and discarding others. It is the fallibilist claim to general validity, or even objectivity, that seems to be difficult to defend in an epistemological sense. This puts fallibilism in a difficult position. As a fallibilist, I might accept that the pretence of general validity is indefensible, but still would like to hold on to the fallibilist justification model – because of the practical relevance connected with it, for instance. This acceptation, however, immediately transforms the fallibilist into a relativist, because he – since he abandons the claim to general validity – must accept that alternative justificatory systems (might) exist, with other epistemological beliefs as their supportive basis. This option will be utterly unacceptable to most fallibilists, as from the very beginning the entire fallibilist project was an attempt to uphold the notion of general validity, despite the inevitability of epistemic uncertainty. This leaves the fallibilist with one other option. In spite of the impossibility of defending general validity in epistemological sense, he might hold on to his beliefs. In that case, the fallibilist – quite bluntly – declares some beliefs to be generally valid, for instance because he sees no reason as to why it should not be the best possible beliefs available. This fallibilist probably assumes that if a better alternative were available, people would recognize it as such, and subsequently acknowledge it as the new best possible alternative. The problem of this position, however, is that it might rest upon an overestimation of humans' abilities to be inspired by new ideas. We can well imagine that, for the assessment of beliefs, we are far more dependent on the beliefs we are already familiar with – because we were brought up with them, for instance – so that we are never able to actually seriously consider any other, substantially different beliefs. Furthermore, the present and past have shown that people reaching agreement on the acknowledgement of epistemological beliefs is exceptional rather than common. In this respect, this fallibilist runs the risk of ethnocentrism, because beliefs in line with the basic belief he is more familiar with will sooner appear to be valuable and acceptable, whereas deviating beliefs will sooner appear to be unacceptable, or even abject; which may lead to a structural exclusion of whatever is 'unfamiliar'. It seems the fallibilist, therefore, is facing a choice: relativism or the risk of ethnocentrism, neither of which seems very appealing, especially in light of their epistemological ambitions.

4. The search for a possible alternative

Although fallibilism offers solutions on some points to the problems that ensued from a traditional, classic foundationalist approach to epistemo-logy, this approach itself also raises a number of questions. At this point in my research, I am led to the conclusion that it seems difficult to reconcile the acknowledgement of the inevitable uncertainty that goes hand in hand with the justification of knowledge claims, with the general validity of knowledge claims on the one hand, and the attribution of epistemological privilege to justificatory grounds, on the other hand. Attempts to nevertheless hold on to it, run the risk of structurally excluding any conflicting perspectives. In my search for the possibilities of an epistemological approach that leaves room for uncertainty, it appears to be necessary to look further. Against the background of the questions raised in my account of fallibilism, I might now be inclined to put back on the track of the radical contextualist epistemological approach as proposed by the authors in chapter two. After all, this approach abandons the claim to general validity, resists the explicit attribution of epistemological privilege and, finally, offers a possibility to deal with the threat of barring the unfamiliar. However, at this point in my research, it is far too early to choose this road. Given the problems arising from the attribution of epistemological privilege and the ensuing concept of a vertically ordered body of knowledge, it does, however, seem interesting to at least explore the notion of a horizontally structured justification model, which is also suggested in antifoundationalism.

Let me first examine what it actually means to say that justification should be regarded as horizontally structured. Generally, it means that claims and their underlying arguments are part of a network of mutually related claims, referred to as the justification context. Regarding this network as horizontal stems from the conviction that it is epistemologically impossible to arrive at a hierarchical order of beliefs within this network. That is to say: none of the claims within this network are epistemologically privileged, so that any claim may be used to justify another claim, but may itself be in need of justification at any other moment. Justification is still possible in such a concept, but it will take on a different nature. The claim that slavery is reprehensible, for instance, can still be used to condemn incidents of people being forced to work with limited freedom. However, the functioning of the argument is now not ascribed to its greater certainty or objective, or general validity, but can be accredited to the position of both beliefs ('slavery is reprehensible' and 'the incident is morally incorrect') within a more extensive network of claims that reinforce one another. The arguments, therefore, serve to clarify that the condemnation of this incident fits in a wider network of beliefs. One of the implications is that a claim may be regarded as justified within the one justification context, whereas it may be labeled as nonsensical in another. We can illustrate this by means of an example borrowed from Elgin (1996, p. 213). At first glance, the statement that 'the moon ate the sun' seems irrational to me. On second glance, however, the statement made by someone from a different cultural background, is found to be embedded in a network of beliefs, presuppositions and implications that make it appear rational after all. This – unfamiliar – network apparently does not include our usual presupposition that an eater – in this case, the moon – should be a living organism.

The functioning of arguments does, so it seems, not necessarily have to be derived from the greater 'certainty' or general validity that we ascribe to them, and that, instead of a vertical justification model, we can also use a horizontal model to clarify what it means to be justified to make a claim. However, to many, the picture of a horizontal model that I have painted thus far will be far from persuasive. Since we still seem to be dependent on justification if we want to make a claim, the threat of complete arbitrariness appears to be diverted. However, does this solution not push us directly towards a form of relativism? Do we have to accept the explanation of someone from a different cultural background, because that explanation fits one's 'network', like 'our' statements fit ours? In that case, a horizontal, contextualist approach to justification only seems to cause perhaps even bigger problems. Indeed, we might be better off with fallibilism, for fallibilists at least do not take the beliefs that substantiate their claims for granted. They continue to ask critical questions. Therefore, the task remains for me to find out in the next chapters whether a contextualist approach to justification is able to outweigh the problems discussed here. To do so, it is important to first gain insight into our exact understanding of the nature and functioning of justification contexts.

Rorty is an author who epistemologically explores the notion of a horizontal justificatory scheme, or of the contextuality of knowledge, without lapsing into relativism, and who moreover discusses the related topic of 'exclusion'. They come together in Rorty's views on 'irony', a concept that is relevant anyway, because philosophy uses it to understand how the inevitable dubitability of our basic beliefs can be dealt with constructively. This is the reason for further investigating the philosophical usability of the concept of irony, and also to extensively discuss the elaboration of that concept by Rorty in the next chapter.

5. References

Adler, J. E. (2003). Knowledge, truth, and learning, in R. Curren (Ed.), *A companion to the philosophy of education* (pp. 206-217). Malden: Wiley-Blackwell.

Baynes, K., Bohman, J. & McCarthy, T. (1987). General introduction, in K. Baynes, J. Bohman & T. McCarthy (Eds.), *After philosophy: End or Transformation?* (pp. 1-18). Cambridge, Massachusetts: The MIT Press.

Blake, N., Smeyers, P., Smith, R. & Standish, P. (1998). *Thinking again, education after postmodernism.* Westport: Bergin & Garvey.

Carr, D. (1998). Moral education and the objectivity of values, in: David Carr (ed.), *Education, knowledge and truth: beyond the postmodern impasse* (pp. 114-128). London: Routledge.

Carr, D. (2000). Moral formation, cultural attachment or social control: what's the point of values education? *Educational Theory, 50*(1), 49-62.

Dancy, J. (1985). *Introduction to contemporary epistemology.* Oxford: Blackwell publishers.

Elgin, C. Z. (1996). *Considered judgment.* Princeton: Princeton University Press.

Rorty, R. (1989). *Philosophy and the mirror of nature.* Princeton: Princeton University Press.

Siegel, H. (1990). *Educating reason. Rationality, critical thinking and education.* New York: Routledge.

Siegel, H. (1997). *Rationality redeemed? Further dialogues on an educational ideal.* New York: Routledge.

Siegel, H. (1998). Knowledge, truth and education, in: David Carr (ed.) *Education, knowledge and truth: beyond the postmodern impasse* (pp. 19-35). London: Routledge.

Williams, M. (2001). *Problems of knowledge.* New York: Oxford University Press

IV
THE FRUITS OF IRONY: GAINING INSIGHT INTO HOW WE MAKE MEANING OF THE WORLD[7]

1. Introduction

The impossibility of redeeming the original ambitions of metaphysics and epistemology is hardly contested these days, which led Hilary Putnam to the conclusion that "philosophical problems are unsolvable" (Putnam, 1990, P. 19). Awareness of this ultimate insolvability is reflected in recent discussions in philosophy of education, which pay a great deal of attention to the resulting necessity of reformulating the ambitions and pretensions of this discipline. Against this background, philosophy of education demonstrates a tendency to accept the ultimate uncertainty of philosophical knowledge, and to replace the primacy of epistemology - including its traditional promises of certainty - with the primacy of engagement, thus severely restricting the pretensions of philosophical claims (cf. Van Goor et al., 2004). However, to accept that one is not able to solve the problem of what the world is really like does not dissuade philosophers from trying to increase their understanding of how we come to know it and make sense of it.

Throughout the history of philosophy, one remarkable way of dealing with various kinds of uncertainty has been to take refuge in 'irony' as a way to continue philosophical research in conditions of insolvability. As Claire Colebrook (2002, pp. 2-3) formulates it: "Irony can be seen as a particular technique that reflects on our conditions of making meaning of the world ... Irony takes those terms that seem to be foundational and opens them up for question". Not giving up on the ambition of making progress in philosophy, these irony-based approaches try to exploit the very uncertainty of philosophical issues to further philosophical understanding. In doing so, the use of irony aspires to convert philosophy's seeming powerlessness into its opposite. Through the ages, the interpretation of irony as a philosophical tool has taken different shapes. Against the background of philosophical insolvability as it is perceived today. I will investigate what gains various philosophers have expected from the use of irony, and I will consider the potential relevance of these gains for philosophy of education. To that end, I will first briefly discuss a few influential historical interpretations of irony as a philosophical tool, before concentrating on two recent efforts to make productive use of irony to gain insight into how we make meaning of the world. These recent

---

7   Published as: Van Goor, R. & Heyting, F. (2006). The fruits of irony: Gaining insights into how we make meaning of the world, *Studies in Philosophy of Education, 25*(6), 479-496 (printed with permission)

interpretations share the idea that no conclusive insights are to be expected, because they all deny the possibility of knowing what the world is apart from any meaning we put on it. However, in other respects these interpretations of irony are based on different theories of meaning-making. Against this background. I develop a third interpretation of irony using yet another theory of meaning-making. I will subsequently discuss the philosophical merits of these three interpretations and the different kinds of insight they can lead to in philosophy of education.

2. Irony as a philosophical tool

According to Gregory Vlastos, all uses of irony in philosophy ultimately seem inspired by Socrates, whose use of the concept has been entrenched in every language of the Western world since Cicero. This meaning reads: "expressing what we mean by saying something contrary to it" (Vlastos, 1991, p. 43). In its most simple and banal forms, ironic expressions are used for humor or mockery. Vlastos mentions the example of a visitor who, arriving in the midst of a downpour, remarks: "What fine weather you are having here". In this example, listeners won't have any trouble understanding that the visitor means the contrary of what he says (Vlastos, 1991, p. 21). In other cases, it does not seem so easy. For example, a teacher who says to a blundering student: "you are brilliant today", may convey his intention to criticize the student for poor performance effectively, but the student will remain in the dark about the nature of his mistakes. As Vlastos (1991, p. 22) expresses it: "He has been handed a riddle and left to solve it for himself". It is this 'riddling' effect that is characteristic of philosophical uses of irony. It furthers insight by evoking questions rather than producing answers.

Socrates was the first to use this kind of irony as a tool in philosophy. Though the question to what ends Socrates introduced this method is much disputed, most authors agree that one of those ends was educational (cf. Gulley, 1968; Vasiliou, 2002; Vlastos, 1991). In his predominant use of irony, "Socrates never directly tells an interlocutor that the interlocutor does not know something. Rather, he operates on the standing assumption that the interlocutor's avowal is correct, and he proceeds to draw out its ramifications" (Vasiliou, 2002, p. 220). In doing so, Socrates lures his interlocutor into a position where he has to reconsider his 'conceit of knowledge' (ibid.), without suggesting any answers. This does not mean Socrates does not care whether people have answers, but, as Vlastos (1991, p. 44) puts it: "he cares more for something else: that if you are to come to the truth, it must be by yourself for yourself".

In his discussion of the educational relevance of Socratic irony, Alven Neiman stresses the importance of recognizing uncertainty, making irony "a means of thinking, of inquiry, in a world freed of absolutes" (Neiman, 1991, pp. 372-373). Paul Smeyers adds another dimension by suggesting that Plato also made use of irony on a more abstract level. By his specific way of presenting Socrates' way of manipulating his inter-locutors, Plato effectively advocated an idea of the art of living: "of developing one's own self and that there are, insofar as virtue is con-cerned, no teachers", while at the same time suggesting that there are no answers (Smeyers, 2005, p. 176). Vlastos (1991, p. 44) implies a similar intention saying: "The concept of moral autonomy never surfaces in Plato's Socratic dialogues - which does not keep it from being the deepest thing in their Socrates, the strongest of his moral concerns". This educa-tional interpretation of Socrates' irony, in which refraining from any answers is considered crucial for developing an authentic, questioning attitude, does not necessarily imply that Socrates did not have any answers or that he thought no answers were possible. In practicing his elenchus - his technique of questioning a position by deriving doubt-inducing consequences from it - he could have feigned ignorance just to motivate his interlocutor to think for himself. Opinions differ on this issue, which makes it hard to determine in what sense Socrates intended to use irony as a philosophical method for gaining insight under conditions of uncer-tainty. Norman Gulley mentions some reasons for understanding Socrates' irony as an indication that his ignorance was not feigned but sincere. First, Plato presents "the elenchus as such a dominant feature of Socrates' arguments that it triumphs over all attempts to reach a positive con-clusion". And discussions are not only inconclusive, in addition "we find Socrates declaring his despair of ever reaching any definite truth", complaining "that all propositions seem to be shifting and transitory, so that they 'run away' from any systematic attempt to substantiate their truth" (Gulley, 1968, p. 67). Nevertheless, Gulley concludes that all this does not bespeak skepticism on Socrates' side, but rather reflects a specific attitude of Plato towards the thinking of Socrates (an attitude charac-teristic of the early dialogues but changing in the later ones) (Gulley, 1968, p. 72 ff.). Vlastos (1991, p. 42), on the other hand, denies that Socrates allows himself "deceit as a debating tactic", whether educa-tionally inspired or not. He designates Socrates as the one who initiated the meaning of 'irony' as a sincere method for evoking insight - an inter-pretation that Iakovos Vasiliou (2002. p. 221) in turn denies.

It may not be a foregone conclusion exactly to what uses Socrates put his irony, but more recent supporters of the method leave less room for doubt. For example, Friedrich von Schlegel (1772-1829) considers irony a method for confronting the limits of what can be represented, without any promise of ever getting a grip on it. In his view, the 'essence' can only be understood as lying beyond what can be said in words. Consequently, the yields of irony in Schlegel's interpretation should be understood not as "a position, form of life, or personality, but ... in what lay beyond any specific position" (Colebrook, 2002, p. 128). Schlegel's irony was meant to gain insight while being aware that the philosophical problem at hand would not be solved. Unlike Immanuel Kant, Schlegel held the opinion that the preconditions for what can be said in words cannot be formulated, because such transcendental preconditions would only induce an infinite regression of asking for preconditions for identifying preconditions as such, et cetera - resulting in what he called a 'transcendental buffoonery' (Comstock, 1987). Therefore, irony leads to an alternating process of saying one thing and recognizing its thorough and necessary banality (Colebrook, 2002, p. 132). In this way Schlegel's interpretation of irony offers a way of developing understanding while relativizing the temptation to answer all questions (cf. Bransen, 1991, p. 172).

Schlegel's conception of irony is reminiscent of Socrates' in that it questions every proposition while at the same time being unable to avoid adducing one - and vice versa. Søren Kierkegaard's interpretation of irony as a philosophical tool seems similar but is more reminiscent of the educational interpretation of Socrates' irony in some respects. In Kierkegaard's interpretation, irony refers to the insurmountable opposition between 'essence' and 'phenomenon'. According to him, the use of irony makes that "the movement is continually in the opposite direction" (Kierkegaard, 1989, p. 257), - an idea also used by Schlegel. In Kierkegaard's discussion of the meaning he attaches to this movement, he especially stresses the "subjective pleasure as the subject frees himself by means of irony from the restraint in which the continuity of life's conditions holds him" (Kierkegaard, 1989, pp. 255-256).

Both dimensions of irony - one pertaining to the self, the other to the claims one is able to substantiate - resurface as a tool in philosophy in the two recent interpretations of irony discuss below. Both interpretations were introduced as a way to tackle the problem of how people give meaning to the world while sharing a contemporary perception of philosophical insolvability as the condition that we are not able to separate 'the world' from the meanings we give to it. The first was formulated by Jan Bransen (1991), who was inspired by Schlegel, and the second by Richard Rorty (1989). Rorty's interpretation of irony is reminiscent of Kierkegaard to the extent that he attaches great importance to the creation and recreation of the self. Though both authors share the view of insolvability as mentioned above, they tackle the problem along different lines. This becomes visible in the specific opposites their respective interpretations of irony imply. They define the opposites between which the ironist philosopher is thought to be alternating differently, which results in a different view of the fruits of irony for understanding the process of meaning-making.

3. Irony and the opposition of conceptual perspective and presupposed reality

Bransen approaches the insoluble problem of how we make meaning of the world most directly from the perspective of the two parties that seem primarily involved in the process: the meaning-making human on the one hand, and the world he makes meaning of on the other. Consequently, Bransen's interpretation of irony involves an alternating process between the opposites of understanding statements as resulting from meaning-making by identifying a - presupposed - object in external reality, and understanding statements as resulting from meaning-making as expressing the conceptual perspective of the meaning- making subject. He illustrates this opposition with everyday words like 'parent' or 'child'. Though we assume that these concepts refer to certain persons who are supposed to exist independent of our conceptual constructions, we cannot avoid the conclusion that parents and children "are what they are because of the way we think about them" (Bransen, 2004, p. 13, trans. authors).

To be able to understand any statement as meaningful - as an informative and not arbitrary enunciation - we have to assume that it is not purely the product of a construction that can be reduced to the conceptual frame of the meaning-making subject. In order to be understood as meaningful, a statement must at the same time be understood as related to a possible reality of an object outside of our constructions about which the statement says something (Bransen, 1991, p. 108). The central problem - that is also responsible for the insolvability of philosophical problems - is that we are unable to separate these 'conceptual' and 'objective' components of meaning-making As a consequence, it will be impossible to get a grip on either component. The unavoidability of these two components combined with the fact that it is impossible to separate them constitutes the basic principle of Bransen's use of irony. As Bransen formulates it, any process of meaning-making that lies at the basis of making a statement can be characterised in terms of finding (the supposed objective component) and at the same time in terms of making (the subjective, conceptual component). Any attempt to fathom this process will inevitably confront us with the "antinomy of thought" (Bransen, 1991, p. 95), because we can only understand the objects of thought as simultaneously determined by thinking itself and by something outside thinking. Bransen develops this idea building on Salomon Maimon, who introduced it in his critique of Kant. Kant thought he had managed to isolate the subjective component by formulating his so-called 'pure concepts of the understanding'. In his critique, Maimon drew attention to the circularity of Kant's argumentation by maintaining that one could only say something about the human faculty of cognition - including the pure concepts of the understanding - by simultaneously making use of this very faculty (cf. Nelson, 1971, p. 30 ff.). This also implies that the separation of the object- and subject-related dimensions of any statement will keep slipping through our fingers, as Bransen explains, leaving the process of meaning-making itself in the dark. According to Bransen, meaning-making cannot be understood purely empirically - as if it were a product of 'reality' alone - nor can it be understood solely as a product of a conceptual perspective - as constructivists sometimes claim. Rejecting both empiricism and constructivism seems to leave only the option of skepticism (the position Maimon chose), and accepting the unfathomableness of all making meaning of the world. Bransen recognizes the problem but he does not see it as the defeat of philosophy. On the contrary, he considers this antinomy a main source of insight for philosophy: even if the process of trying to understand meaning-making does not result in any final answers, the typical course of this effort - which he calls 'irony' - still offers a kind of understanding that can motivate the philosopher to keep asking questions and continue his investigations (Bransen, 1991, p. 175).

Though neither empiricism nor constructivism suffices to understand any instance of meaning-making, Bransen's use of irony incorporates both. If we take a dogmatic constructivist view, acting as if claiming that meaning were only the result of making, we will inevitably be confronted with the object-related component of meaning-making - if only because the resulting statement will not fully coincide with whatever it describes - and we will inevitably be forced into the negation of this claim. For example, if we consider the statement 'this is a child' as purely the result of making, it would be almost impossible to explain what is meant by 'this' without assuming something outside this statement to which this statement refers and about which it says something. In addition, the question whether the statement 'this is a child' should be considered 'right' in the sense of 'corresponding to the world' is highly unlikely to arise as long as we see it purely as a result of making. The statement would simply reflect the conceptual perspective of its maker. Any difference of opinion on this issue would either be pointless - reducing the statement to a mere expression - or require an appeal to some external reality to settle it - thus abandoning the constructivist position. Consequently, if we consider any informative statement from a 'pure' constructivist perspective, the idea that there must be something 'out there' to which the statement refers, will force itself upon us. In short, taking a dogmatic-constructivist position will evoke its empiricist opposite according to Bransen.

Starting from the alternative dogmatic-empiricist position and acting as if meaning were only the result of finding, we will equally inevitably be confronted with its constructivist opposite, with the dimension of the meaning that is made, i.e. the conceptual perspective of the subject. To consider the statement 'this is a child' as purely the result of finding would in fact exclude all differences of opinion, and certainly all discussion about alternative definitions. Acting as if meaning were only the result of finding would essentially mean that alternative definitions of 'child' are erroneous, because finding alone cannot offer any instruments for solving the problem of a variety of definitions contending for the designation 'right'. Even if we define explicitly what to look for in 'finding' a child, we would be confronted with the necessity of conceptually mediated access to reality. Consequently, acting as if meaning were the result of finding inevitably confronts us with the indispensability of a conceptual perspective, i.e. with the dimension of making. This use of irony, then, owes its effectiveness to the phenomenon that taking one position inescapably leads to its opposite. Bransen's interpretation of irony causes an alternating movement between dogmatic empiricism and dogmatic constructivism; it is the inconclusive nature of this very movement which furthers insight into the phenomenon of meaning-making - without ever reaching a conclusive insight. Bransen does not claim to separate both components of meaning-making, and he explicitly rejects the idea that his approach, however indirectly, promises any view of external reality 'as it is', or isolates the subjective contribution to meaning. Rather, his use of irony induces "a consciousness in motion, such that a deliberate support for either one of the convictions uncovers an unintended support for the other one" (Bransen. 1991, p. 171). The gain of this approach consists in an insight into the unfathomable nature of meaning-making that results from this 'consciousness in motion'. The moment of irony "uncovers, via an absurdity (i.e. through irony) the plausibility of the opposing account- (Bransen, 1991, p. 182). As a consequence, this use of irony does not result in solutions, but stimulates further questions. The insight that results from this procedure will neither reveal the 'real' nature of what a child 'is', nor the 'right' perspective to conceptualize it. In this interpretation irony implies bringing our insight in motion, urging us to alternately concentrate on the conceptual frames that model our statements and on the presupposed reality about which these statements are meant to inform us. In Bransen's words: "The point is, in other words, that, in order to ask the right questions, we have to be sensitive to what is beyond our grasp" (Bransen, 1991, p. 31n). To be clear: Bransen does not claim that each statement is the result of a conceptual perspective and a supposed reality, but he does maintain that his theory - including his conception of irony as a philosophical tool - can provide the most plausible explanation of our meaning-making practice to date. Though we have no means to achieve a decisive view of the 'real' object or even of its 'existence' we have to presuppose it in order to be able to explain the informative nature of statements. according to Bransen. We act "as if there was a 'transcendent object' we try to make sense of", as he puts it (Bransen, 1991, p. 3 In). However, this explanation of making meaning of the world also raises questions that will affect the interpretation of the philosophical use of irony.

4. Irony and the opposition between vocabularies and aims

Bransen's interpretation of irony as a philosophical tool presupposes the distinction between conceptual perspectives and supposed external reality as the basic poles of meaning-making. However, not all philosophers think such a distinction can be meaningfully made. According to Nelson Goodman, for example, all we have at our disposal are 'versions' and "all that can be done to comply with the demand to say what the versions are versions of is to give another version" (Goodman, 1989, p. 83). In other words: introducing (the supposition of) an external object actually means introducing another 'version', another statement from another perspective and nothing else. However, if we abandon the view that the distinction between conceptual frame and presupposed reality is necessary for understanding any statement as meaningful, we must also abandon Bransen's interpretation of irony and its yield for philosophy. Rorty's interpretation of irony - which does not utilize the disputed distinction - might offer a suitable alternative; it would imply an alternative way of furthering insight into the process of making meaning of the world.

Just like Goodman, Rorty argues that introducing the presupposition of an 'external' reality will not help us understand the nature of meaning-making, because "the world is out there, but descriptions of the world are not" (Rorty, 1989, p. 5). Consequently, assuming the 'world' causes statements does not help to explain how those statements are made. Rorty suggests the role of vocabularies or language games as an alternative. According to Rorty, we are inclined to forget that single statements belong to vocabularies that imply criteria for formulating and accepting statements. When making a statement, we implicitly accept the criteria of the vocabulary in which the statement is embedded. Against this background it becomes difficult to think of the world in terms of deciding between vocabularies (Rorty, 1989, p. 5). Vocabularies as Rorty understands them, do not depict the world or even selected parts of it, nor do they express the state of the subject. If they did, we could combine them to form a more complete picture of the world or the subject - but trying to do so will only result in absurd questions like "What is the place of consciousness in a world of molecules?" (Rorty, 1989, p. 11).

If it is meaningless to separate the world from what we say about it, and if vocabularies are all we have access to, we will have to concentrate on the use of vocabularies to understand the process of making meaning. As Rorty sees it, vocabularies do not 'represent' the world, and they are not instruments for pure subjective expression either. Rather, he considers vocabularies as tools for coping. For example, using a mentalistic vocabulary, e.g. by saying that people have a 'mind' "is just to say that, for some purposes, it will pay to think of them as having beliefs and desires" (Rorty, 1989, p. 15). However, according to Rorty this analogy between vocabularies and tools falls short in one respect: whereas the craftsman knows "what job he needs to do before picking or inventing tools with which to do it", we do not know in advance what aims we want our vocabularies to be used or designed for (Rorty, 1989, p. 12- 13). Inspired by Donald Davidson, Rorty develops his theory of vocabularies in the basis of this amended Wittgensteinian view of language as a coping tool (Rorty, 1989, p. 13-15). This conception of vocabularies - between which the ironist philosopher can alternate without ever identifying the 'one and only right' vocabulary - lies at the basis of Rorty's interpretation of irony and how it can further insight into the process of making meaning of the world. Rorty presents his interpretation of the fruits of irony primarily with respect to persons and their 'final vocabularies', in which they justify their actions and beliefs and formulate their deepest doubts and highest hopes. When asked for further justification, "their user has no noncircular argumentative recourse" (Rorty, 1989, p. 73), which means that at this point a user is possibly aware of being in a vocabulary with its characteristic limitations and possibilities. In Rorty's view, an ironist is someone who is inclined to doubt his or her current final vocabulary because (s)he knows of the existence of alternative ones, while also being aware of the impossibility of confirming or dissolving these doubts in any of those vocabularies (ibid.). In this sense an ironist is aware of the 'contingency of the self (Rorty, 1989, p. 74). Here Rorty' s interpretation reminds us of Kierkegaard, who also stressed the perspective of the subject who frees himself from conventional restraints.

Though introduced with respect to the self and the 'final' vocabulary that is used to present it, this interpretation of the philosophical use of irony can be applied to any vocabulary and whatever it is meant to accomplish. Rorty challenges the tendency to think of the world and of the self as possessing an intrinsic nature that can be caught in language. He

75

contests that "there is some relation called 'fitting the world' or 'expressing the real nature of the self which can be possessed or lacked by vocabularies-as-wholes", and the "temptation to privilege some one among the many languages in which we habitually describe the world or ourselves" (Rorty, 1989, p. 6). In other words, irony not only furthers insight into the vocabulary-bound process of making meaning of the self, but it also furthers insight into the vocabulary-bound process of making meaning of the world. In both cases, the irony consists in the moment where the analysis of the tenability of a statement - understood as an instrument for coping - gives cause to a confrontation with the vocabulary that made this statement possible in the first place, as well as with the criteria for its acceptability. Irony as Rorty sees it involves radical doubts about the use of any vocabulary, because of the awareness that it will favour certain aims while excluding others. At the same time Rorty's interpretation of irony implies "seeing the choice between vocabularies as made neither within a neutral and universal metavocabulary nor by an attempt to fight one's way past appearances to the real, but simply by playing the new off against the old" (Rorty, 1989, p. 73). This kind of irony will result in a never-ending process of re-description, of re-orientation in the world by using or creating alternative vocabularies, while never being able to prove one of them 'best'. Like Bransen's approach of irony, Rorty's does not result in any kind of certainty. Instead, both approaches enhance sensitivity to the inabilities and restrictions of our practices of meaning-making, sensitivity to what keeps escaping our grip. However, both authors have different interpretations of what it is exactly that is presumed to escape. In Bransen's interpretation it is the intangible interplay between conceptual perspective and the world; in Rorty's interpretation it is the irreducible interplay between vocabularies and the immanent aims they appear to favor in the process of coping. For example, with respect to explanations of what a child 'is', Rorty's interpretation of irony as a philosophical tool will draw attention to the specific vocabulary each explanation was embedded in, and to the specific coping-related descriptions it allows for, resulting in an awareness of the contingency of each. Thinking of the oppositional conceptions of the child propagated by 'care-takers' and 'liberationists' (Archard, 1993), the care-takers' conception of the child as vulnerable and dependent will reveal a vocabulary primarily aimed at protection and guidance, whereas the 'liberationists' conception of the child as an active agent will reveal a vocabulary that primarily allows for dealing with children in terms of stimulation and facilitation. To some extent this approach is reminiscent of the prominent attention for 'counternarratives' (Peters & Lankshear, 1996) and 'counterpractices' (Biesta, 1998) in current critical educational theory.

In order to draw attention to the idea that views of educational issues do exclude certain groups and viewpoints, these approaches point out alternative views and descriptions, without claiming to offer a perspective that will not exclude any people or viewpoints. While excluding effects - or at least restrictions - can be brought to light and made a subject of discussion, they cannot be avoided. Biesta (1998, p. 507): "A counter-practice should not be designed out of an arrogance that it will be better ... than what exists. A counter-practice is only different". In a similar way Rorty's view of irony causes uncertainty and divergence instead of certainty and convergence. Though this may seem a loss to some, in Rorty's perspective - as in that of his abovementioned educa-tional colleagues - it is quite the opposite, again, because it stimulates sensitivity to the unfamiliar, a process which entails the realization of an ideal in itself.

Rorty's interpretation of the use of irony draws attention to the impossibility of conclusively deciding what is 'objectively' the 'best' vocabulary; it only results in a proliferation of vocabularies. This proli-feration represents its main proceeds, because it stimulates doubt, reflection, and discussion. Against this background Rorty speaks of 'enlarging the canon', which should replace "the attempt by moral philosophers to bring commonly accepted moral intuitions about parti-cular cases into equilibrium with commonly accepted moral principles" (Rorty, 1989, p. 81). Bransen, on the other hand, qualifies the proceeds of Rorty's approach as a kind of fooling, an arbitrary juggling with perspec-tives (Bransen, 1992, P. 173). However, one may wonder whether this verdict does not erroneously identify Rorty's position with that of Bransen's own 'dogmatic constructivism'. Rorty does consider statements - resulting from man-made vocabularies - as 'made', but contrary to Bransen's dogmatic constructivist, Rorty does not consider them arbitrary conceptual creations. As Rorty formulates it: "The rea-lisation that the world does not tell us what language games to play should not ... lead us to say that a decision about which to play is arbitrary, nor to say that it is the expression of something deep within us. The moral is not that objective criteria for choice of vocabulary are to be replaced with subject-tive criteria, reason with will or feeling" (Rorty, 1989, p. 6). Rorty views vocabularies as tools, as instruments for producing statements that help us set aims and strive for them, as instruments that are developed and abandoned in the course of cultural history (Rorty, 1989, pp. 16-17). There seems to be no reason why supposing a reality, as Bransen does, is less arbitrary than supposing a vocabulary. However, on closer inspection Rorty's interpretation of irony, which is based on his theory of vocabu-laries, still raises some questions.

5. Irony and the opposition of context and renewing content

Robert Brandom issued a fundamental critique of Rorty 's approach of vocabularies. Brandom agrees with Rorty that the form, content, and acceptability of our statements are regulated by vocabularies, by "shared norms that antecedently govern the concepts one deploys in making such a claim" (Brandom, 2000b, p. 176). And just like Rorty. Brandom regards vocabularies as 'tools'. However, Brandom does not share Rorty's view of the nature and the use of vocabularies in all respects. According to Rorty, practically anyone can handle a diversity of vocabularies - this is the background of his idea that irony will be fruitful not only in a philoso-phical setting, but in a private or social setting as well. However, not everyone can create a new vocabulary. As Rorty sees it, the creation of a new vocabulary requires the genius of what he calls 'a poet' (Rorty, 1989, p. 43). Brandom disagrees. He considers the development of new vocabu-laries an everyday phenomenon. In his view new vocabularies should not be considered the result of one major creative effort of an exceptionally gifted individual, but rather the result of a continuous, piecemeal process in which every language user participates. According to Brandom, voca-bularies are changed and renewed simply by using them in the process of communication. "To use a vocabulary is to change it. This is what distin-guishes vocabularies from other tools" (Brandom, 2000b, p. 177). As a conse-quence, Brandom's view on the way vocabularies can be inter-related also differs from Rorty's. He does not consider them separate and incompatible, and in his view Rorty overrates the novelty of new vocabu-laries. According to Brandom vocabularies result from each other in use.

In a reaction Rorty agrees with Brandom: "I have been in danger of overromanticizing novelty by suggesting that great geniuses can just create a new vocabulary ex nihilo. I should be content to admit that geniuses can never do more than invent some variations on old themes, give the language of the tribe a few new twists" (Rorty, 2000, p. 188). This way of putting things also has consequences for how irony can be understood to function as a tool in philosophy. If we admit that voca-bularies are changed and renewed by using them, and that any statement expresses a conventional linguistic practice - a vocabulary - and renews it at the same time, it will be difficult to sharply demarcate the different vocabularies. Consequently, it would be inconsistent to consider any spe-cific statement an expression of one specific vocabulary while excluding all others.

Rorty's appeal to the regulating influence of 'vocabularies' will result in a similar problem as Bransen's appeal to the regulating influence of a supposed 'object'. As we can only have access to concrete, linguis-tically expressed statements, the vocabularies these formulated statements are supposed to stem from are just as inaccessible and 'external' as presupposed objects in reality. It does help as little to distinguish descripttion and vocabulary from each other, as it does to distinguish a presupposed object from the conceptual perspective that is used to describe it. A 'new' vocabulary can only be distinguished from an 'old' one by means of a statement that in itself would represent a vocabulary, and it remains unclear what kind of vocabulary that could be.

This critique of Rorty's theory implies the necessity of reformulating his idea of the use of irony. A first step in that direction could be to assume that a statement does not express a specific, separately 'existing' vocabulary, but in a way creates it by being formulated. Brandom acknowledges this by depicting the maker of a statement as someone who communicates, which means: as someone who addresses a public. This emphasis on the communicative function of making statements depicts the speaker as someone who assesses what conventions with respect to meaning-making (what 'vocabulary' in Rorty's terms) his public might share in that particular situation at that particular moment in order to decide what he wants to add to those conventions (his statement). In Brandom's theory, these "shared norms that antecedently govern the concepts one deploys" (Brandom, 2000b, p. 176) take the place of Rorty's vocabularies. Brandom emphasizes that a linguistic expression can only be understandable to a public insofar as it follows shared norms, and it can only be informative insofar as it adds to this convention. In his words: "Every use of a vocabulary, every application of a concept in making a claim, both is answerable to norms implicit in communal practice - its public dimension, apart from which it cannot mean anything (though it can cause something) - and transforms those norms by its novelty - its private dimension apart from which it does not formulate a belief, plan or purpose worth expressing" (Brandom, 2000b, p. 179).

In view of a revised interpretation of the philosophical use of irony, I would add at this point that the novelty of a statement - what it will add or change - cannot be determined in advance by comparing the statement to the set of publicly shared norms themselves, but only by comparing the statement to the norms a speaker thinks his audience will share (or, from the perspective of the audience, to the norms a listener supposes a speaker ascribes to his audience). The publicly shared norms themselves should not be imagined separately, independent of the practice of linguistic use. They only exist in the mind of language users: the speaker who attributes them to his public; the audience that attributes them to a speaker (and if speaker and audience attribute differently they will misunderstand each other); and finally, the analyst who attributes them to the statements of subsequent speakers. It should also be clear that such supposedly shared norms do not reside as fixed sets in the minds of those participants; they are constantly updated with every statement that is made or received in the course of communication (Brandom, 2000a, pp. 164-165). I will call such momentarily attributed shared norms the 'communicative context' of a statement (cf. Stalnaker, 1999). Consequently, a communicative context is not to be considered part of the situation as such, it only belongs to the situation as participants in communication define it at that specific moment (while leaving open the possibility that they define it differently).

Against this background an interpretation of the use of irony has to be developed in terms of an opposition between (supposedly) shared norms, i.e. the communicative context and intended addition to that context, and not in terms of the opposition between vocabulary and aim (way of coping), as Rorty's interpretation of the use of irony implies. This third version of understanding the use of irony is based on the idea that the informative value of any statement should be understood by viewing it as an addition to a communicative context, and - in case of its acceptance - as a renewal of this communicative context. Trying to understand the informative nature of a statement, my ironist would find himself in an alternating movement between communicative contexts the communicator is supposed to have had in mind and the related renewing of the contents of the statement. Like any form of irony, this version will inevitably fail to provide decisive answers, because communicative context and renewing content cannot be accessed separately by any language-user or analyst. Though it seems clear that we will never be able to separate the two components - content and context - of statements, this interpretation of irony will still result in insight into the process of meaning-making and its intangible fluctuation between 'new' and 'old'. The central importance of 'new' and 'old' as well as the concentration on the linguistic dimension of meaning-making, are reminiscent of Rorty's conception of irony as a tool, but my interpretation is also cognate to Bransen's. And although I replaced his opposition of presupposed object and conceptual frame with the opposition of communicative context and renewing content, the motivation of my ironist resembles that of Bransen's. It concerns the unattainable desire to get a grip on both components that are considered to constitute the process of making meaningful statements, rather than the desire towards self-(re)creation which motivated Rorty's ironist. The primary motive of my ironist is the wish to understand the communicative transformations that take place in the intangible interplay between context and content.

The difference between the proceeds resulting from Rorty's interpretation of irony and from my version, that is based on Brandom's theory, can be illustrated with the interpretation of a short oppositional episode in philosophy. Rorty (1989, p. 133 ff.) discusses a debate between Jacques Derrida and John Searle, in which both authors seem to be talking at cross-purposes. Searle reproaches Derrida with not doing justice to the philosophical work of John Austin. In his reply to Searle, Derrida refuses to systematically respond to Searle's points of criticism. Rorty explains this episode as a collision of vocabularies, labeling Derrida's contributions as those of an ironist who "refuses not because he is 'irrational' or 'lost in fantasy', or too dumb to understand what Austin and Searle are up to, but because he is trying to create himself by creating his own language-game ... trying to get a game going which cuts right across the rational-irrational distinction", as would be characteristic of the established philosophical vocabulary. In Rorty's interpretation, Derrida - motivated by the wish for self-creation - is trying to create a new vocabulary. Because this new vocabulary differs from the (established) philosophical vocabulary Searle uses, Searle is unable to understand Derrida's text as a meaningful reaction to Austin. By creating his own language game, Derrida might hope to create an alternative "pattern of linguistic behavior which will tempt the rising generation to adopt it" (Rorty, 1989, p. 9) as an alternative philosophical vocabulary.

My ironist, on the other hand, would understand Searle's critique of Derrida as related to the philosophical communication context Searle implicitly assumed as the context Derrida would have imagined in making his statements about Austin. By saying that Derrida did not do justice to the philosophical work of Austin (and to Searle's critical comments). Searle took this assumed context - which supported his own conclusion that Derrida had made 'mistakes' - for granted. However, ironic concentration on the intangible interplay between content and context could have drawn attention to the possibility that Derrida's texts could also be understood as adding something 'new' to an alternatively imagined context - for example a context of notions about what constitutes 'philosophical' discourse and 'philosophical' plausibility and what does not. Against the background of a context imagined like this, Derrida's texts could be understood as adding something 'new' to the definition of philosophical discourse and its inherent criteria (such as the distinction between 'rational' and 'irrational'). According to this interpretation, irony would draw attention to unavoidably presupposed contexts, making their definition accessible for discussion and transformation within the current discourse. Like other interpretations of irony as a philosophical tool, this approach does not claim to bring us closer to the Truth. This kind of ironic exercise will not reveal the communicative contexts as 'really' imagined by either Derrida or Searle, but it still will further insight by motivating attempts to imagine alternative combinations of presupposed context and informative content. Each renewed turn of irony will stimulate further questions, and the process as a whole will generate an insight into the game of context and renewal.

In my approach the philosophical yield of irony results from the opposition of the (supposed) communicative context and renewing informative content. This approach also suggests a different explanation of the working of counternarratives or counterpractices, not understanding them as alternative vocabularies - as was suggested against the back-ground of Rorty's interpretation of the use of irony - but as potentially renewing predominant communicative conventions. On closer inspection, authors who write about counternarratives and counterpractices give different explanations as well. By stressing mutual incommensurability between "a heterogeneity of different moral language games", Michael Peters and Colin Lankshear (1996, p. 3) seem to argue along the lines of Rorty's idea of playing off different vocabularies against each other. Other authors seem to understand the function of counternarratives more in line with Brandom's theory and my interpretation of the philosophical use of irony.

For example, Gert Biesta stresses that an interpretation of an incommensurable heterogeneity of countenarratives "only makes sense as long as we believe that we occupy a place outside the system from which the system can be viewed" (Biesta, 1998, p. 507). This observation ties in with the problem of distinguishing vocabularies, as I mentioned in my critique against Rorty. In Biesta's interpretation the function of counter-practices should not be understood as playing off different practices against each other, because the "practice of transgression is not meant to overcome limits (not in the least because limits are not only constraining but always also enabling)". This 'enabling' function of limits corresponds to the function of (presupposed) communicative contexts as a prerequisite for making meaningful statements while at the same time limiting their possible range. Henry Giroux makes a similar point. According to him, counternarratives should not be viewed as making use of completely alternative vocabularies, but as attempts by "educators to fashion a critical politics of difference not outside but within a tradition" (Giroux, 1997, p. 152). What Giroux calls 'politics of difference' refers to a philosophical attitude that emphasizes the local, the partial, and the contingent (Giroux, 1997, p. 151). Just like my interpretation of the use of irony, which draws attention to the interrelatedness of the 'new' and the 'old', Giroux considers a counternarrative "a vision of public life which calls for an ongoing interrogation of the past that allows different groups to locate themselves in history while simultaneously struggling to make it" (Giroux, 1997, p. 158).

6. Exemplification and comparison: an occasion for meta-irony

I have discussed three different interpretations of irony as a tool in philosophy, resulting from three different ways of considering statements as expressing 'making meaning of the world'. None of these interpretations allows for a complete understanding of how this meaning- making functions exactly, but all three of them at least claim to enhance insight into the process. Though I did give reasons for developing my own approach on the basis of Brandom's theory, I do not claim to have developed a 'final' view of the use of irony. Before exploring this comparative issue in more depth, I will briefly exemplify each of the three interpretations by applying them to the issue of 'students at risk' and the programs that have been developed to tackle this problem. I will start with a broad definition by Robert Slavin that characterizes 'students at risk' as those students who "are failing to achieve the basic skills necessary for success in school and in life" (Slavin, 1989, p. 5). Though most workers in the field will subscribe to this definition, it appears to be less unequivocal than it may seem at first sight. Each of the three varieties of the use of irony will throw a different light on this issue.

From Bransen's point of view irony will reveal the insolvable ambiguity of how we recognise a student as being 'at risk'. Taking the position of dogmatic empiricist and considering the statement as a result of finding, we act as if the statement this is a student at risk' forces itself upon us as a property of 'the world'. From this position, any difference of opinion about such a statement would be impossible, and if it would occur nevertheless, the dogmatic empiricist position would forbid saying anything that could contribute to its solution. Even trying to explain what to look for in order to recognise a case of 'at risk' would imply the inevitable role of concepts - in this example the role of a definition. Such an approach would also leave the practice of discourse completely unexplained, because the apparent differences of opinion in the field are not treated as insolvable at all. Two of the most influential programmes for 'students at risk' - Success for All and High/Scope - appear to hold different views of what counts as 'necessary basic skills for success'. Whereas Success for All considers academic skills, i.e. skills in the domain of language and cognition, as covering the preconditions for success (Madden et al., 1989; Slavin, 2002; Slavin & Madden, 1989), High-Scope defines a much broader range of skills, including social, emotional, and motor skills, as necessary basic skills for success (Schweinhart Weikart, 1986; Schweinhart & Weikart, 1997; Schweinhart & Weikart, 1998). A next step in Bransens' view of irony would involve considering the possibility that recognition of a student as being 'at risk' would result completely from the conceptual perspective that is taken. However, assuming that position would make designating someone as being 'at risk' a matter to be decided at the discretion of the subject's conceptual perspective. Also, meaningful discussion of the issue would be impossible, because any reference to experience, observations, and research results would be forbidden - and this is precisely what discussants do all the time. However, referring to 'the world' will not settle the issue either, as the position of the dogmatic empiricist demonstrated. The result is a 'consciousness in motion' (Bransen, 1991, p. 171) between the conceptual and empiricist components of statements about 'being at risk'. It will not settle the issue of what being at risk 'is', but it will stimulate further questions and thus help to gain more in-depth insight in the issue.

Putting Rorty's approach of irony into practice would first result in identifying both interpretations of being at risk as originating from two different vocabularies, each allowing for different coping-related descripttions while excluding others. In my example, the concentration on linguistic and cognitive skills in 'Success for All' can be understood as stemming from a vocabulary that primarily relates education to a future position in society, especially a position on the labor market (Stringfield & Land, 2002, p. vii; Winch & Gingell, 2004, p. 6), and that favors aims related to the vocational future of children. This educational vocabulary has a long tradition, as does its opposite, from which the broader approach of necessary skills for future success in 'High/Scope' can be understood. This alternative educational vocabulary aims primarily at "the education of the integral whole child" (Adelman, 2000, p. 103), at becoming a "whole man in his social context" (Rohrs, 1995, p. 12). This vocabulary is often called 'child-centered' or 'progressive' (Brehony, 2000) and pays more attention to liberal aspects of education (Winch & Gingell, 2004, p. 6). After making this first step from the perspective of Rorty's interpretation of the use of irony, the next one follows almost naturally. Once we have recognized the vocabulary-relatedness of being considered 'at risk', it would seem logical to ask whether alternative vocabularies could be possible, and how they could shed light on the way current vocabularies favor certain aims and orientations in education while excluding others. This could draw attention, for example, to what should be considered 'success' and to the possibility that both current vocabularies - no matter how different they are - still represent norms of 'excellence' that are related to dominant cultural groups in society (Margonis, 1992). As menioned before. Rorty's conception of irony will not result in one 'right' vocabulary, but cause a proliferation of vocabularies, an oscillation between considering aims and vocabularies, putting each aim in perspective and furthering understanding of the restrictive nature of each vocabulary.

In conclusion, my interpretation of irony as a tool in philosophy does not relate statements to relatively stable and restrictive vocabularies, but understands them as attempts to amend presupposed contextually accepted standards. in this approach, understanding a statement will require an identification of how it deviates from such standards. For example, the presentation by Robert Slavin and Nancy Madden (1993) of the positive scores for the effectiveness of Success for All can be understood as an attempt to add the desirability of implementing this program in schools to a presumed context in which the priority of linguistic and cognitive skills are accepted standards. However, presentation of these favorable scores can also be understood as an attempt to convince an audience that favors High/Scope to accept Success for All in a presumed context where fighting or preventing any kind of being 'at risk' is an accepted standard. As there is no way of diagnosing the 'real' context - because neither speaker nor audience or analyst can have direct access to the assumptions of the interlocutors-, making as well as understanding any statement involves a (re)construction of content and matching context. In each case the analyzing philosopher can understand the informative content of a statement by reconstructing the part of the context he thinks a speaker assumes and wants to attack. The philosopher now understands the statement as aiming to change this specific part of the context while making use of other aspects of this perceived context in order to be able to make him/herself understood and potentially accepted. Consequently, it depends on the reconstructed context what renewing content is perceived by the audience and what kind of reply would seem appropriate - and each possible reply would testify to a differently changed context. This interpretation of irony as a philosophical tool implies a fluctuation between content and context, potentially furthering an understanding of how one changes the other - although never resulting in a conclusive understandding. Without being able to definitively solve the problem of understandding statements and how we make meaning of the world, it enhances insight into the development of thinking by alternately concentrating on restrictive and renewing dimensions. It helps to make our changing - and at least supposedly - communicative assumptions explicit and bring them up for discussion.

This investigation again raises the question whether the reasons I formulated for preferring my third interpretation of the use of irony can be considered decisive. Is this the ultimate philosophical conception of irony as a philosophical tool? I don't think I can - or want to - substantiate such a claim. Like Bransen's and Rorty's, my interpretation does not - and cannot - claim to overcome the insolvability of philosophical questions in any way. Instead, all three interpretations aim at making insolvability productive without removing it. Pretending to have developed the 'ultimate' explanation of the use of irony would contradict the meaning of 'irony' itself, at least in the definition by Colebrook (2002, pp. 2-3) which I took as my starting point: as a technique for reflecting upon the pre-conditions of making meaning of the world, not to establish the 'right' pre-conditions once and for all, but to be able to bring any reconstructed preconditions up for discussion.

What I can do is put the three versions of the use of irony that figured above to the test of irony as I interpreted it. This would result in a kind of meta-irony which would not only challenge the informative nature and imagined context - including the presupposed shared norms - of Rorty's and Bransen's interpretations, but the informative nature and imagined context of my own interpretation as well. In reply to Brandom's critique Rorty remarks that he had not realized he had devised an ap-proach that in its turn depended on a specific vocabulary, a "vocabulary vocabulary" (Rorty, 2000, p. 188). In his case, the concept of 'vocabulary' obviously functions as a basic assumption he supposes his audience to be able to share with him. A closer look at Bransen's interpretation of the use of irony reveals the supposedly shared basic assumption that statements have a representational nature and also result from a subject-dependent conceptual frame. Finally, my own version seems characterized by the presupposed shared basic assumption that any meaningful statement should be considered part of an ongoing communication, i.e. is formulated to inform others instead of being informative by itself. The irony in all of this is that all three versions appear to depend on a supposedly shared basic assumption - whether seen as communication-related or not - that is both fallible and indispensable at the same time. Without such an as-sumption - whichever one prefers - it would not be possible to define any interpretation of the use and the yield of irony at all. So this is an illus-tration of meta-irony: it consists in playing off the different interpretations of irony as a philosophical tool against each other, bringing them all up for discussion without being able to designate one of them as decisive.

Stimulating new questions and new directions to find them summarizes how all versions of irony make philosophical insolvability productive without ever overcoming it. The importance of the different varieties of irony is that they give us "a new problem instead of an old principle [trans. RvG]" (Bransen, 1992, pp. 179-180). This meta-level of irony also puts the very idea of insolvability in a new light. It makes the idea of a (potentially) 'right' basic assumption suspect, and replaces it with the idea of fallible - though indispensable - basic assumptions. The fruits of irony as discussed here largely consist in revealing such basic assump-tions and bringing them up for discussion, while at the same time being aware that other 'conceptual perspectives', 'vocabularies' or commu-nicative 'contexts' are inevitably at work and kept beyond discussion - at least for the time being. Such procedures will replace hopes for 'the right' basic assumption with attention to the process of making meaning of the world itself.

7. References

Adelman, C. (2000). Over two years, what did Froebel say to Pestalozzi? *History of Education, 29*(2), 103-114.

Archard, D. (1993). *Children: Rights and childhood.* London: Routledge.

Biesta, G. J. J. (1998). Say you want a revolution...Suggestions for the impossible future of critical pedagogy. *Educational Theory, 48*(4), 499-510.

Brandom, R. B. (2000a). *Articulating reasons: An introduction to inferentialism.* Cambridge, MA.: Harvard University Press.

Brandom, R. B. (2000b). Vocabularies of pragmatism: Synthesizing naturalism and historicism, in: R. B. Brandom (Ed.), *Rorty and his critics* (pp. 1-56-183). Malden, MA: Blackwell Publishing.

Bransen, J. (1991). *The antinomy of thought: Maimonian skepticism and the relation between thoughts and objects.* Dordrecht: Kluwer Academic Publishers.

Bransen, J. (1992). *Filosofie & ironie. Fantastische opmerkingen over de toekomst van een traditie.* Kampen: Kok Agora.

Bransen, J. (2004). *Jezelf blijven.* Nijmegen: Katholieke Universiteit Nijmegen.

Brehony, K. J. (2000). Introduction. *History of Education, 29*(2), 97-101.

Colebrook, C. (2002). *Irony in the work of philosophy.* Lincoln: University of Nebraska Press.

Comstock, C. (1987). "Transcendental buffoonery": Irony as process in Schlegel's "Über die Unverstandlichkeit". *Studies in Romanticism, 26*(3), 445-464.

Giroux, H. A. (1997). *Pedagogy and the politics of hope: Theory, culture, and schooling: A critical reader.* Boulder: Westview Press.

Goodman, N. (1989). "Just the facts, ma'am!", in: M. Krausz (Ed.), *Relativism, interpretation and confrontation* (pp. 80-85). Notre Dame: University of Notre Dame Press.

Gulley, N. (1968). *The philosophy of Socrates.* London: MacMillan and co.

Kierkegaard, S. (1989). *The concept of irony; with continual reference to Socrates* (H. V. Hong & E. H. Hong, Trans.). Princeton: Princeton University Press.

Madden, N. A., Slavin, R. E., Karweit, N. L., & Livermon, B. J. (1989).

Restructuring the urban elementary school. *Educational Leadership, 46*(5), 14-18.

Margonis, F. (1992). The cooptation of at risk': Paradoxes of policy criticism. *Teachers College Record, 94*(2), 343-364.

Neiman, A. (1991). Ironic schooling. Socrates, pragmatism and the higher learning, *Educational Theory, 41*(4), 371-384.

Nelson, L. (1971). *Progress and regress in philosophy. From Hume and Kant to Hegel and Fries, Vol. II* (H. Palmer, Trans.; J. Kraft, Ed.). Oxford: Basil Blackwell.

Peters, M., & Lankshear, C. (1996). Postmodern counternarratives, in: H. A. Giroux, C. Lankshear, P. McLaren & M. Peters (Eds.), *Counternarratives - cultural studies and critical pedagogies in postmodern spaces* (pp. 1-39). New York: Routledge.

Putnam, H. (1990). *Realism with a human face* (J. Conant, Ed.). Cambridge, MA: Harvard University Press.

Rohrs, H. (1995). Internationalism and development of progressive education and initial steps towards a world education movement, in: H. Rohrs & V. Lenhart (Eds.), *Progressive education across the continents: A handbook* (pp. 11-27). Farnkfurt am Main: Peter Lang.

Rorty, R. (1989). *Contingency, irony, and solidarity.* Cambridge: Cambridge University Press.

Rorty, R. (2000). Response to Brandom, in: R. B. Brandom (Ed.), *Rorty and his critics* (pp. 183-190). Malden, MA: Blackwell Publishers.

Schweinhart, L. J., & Weikart, D. P. (1986). Early childhood development programs: A public investment opportunity. *Educational leadership, 44*(3), 4-12.

Schweinhart, L. J., & Weikart, D. P. (1997). The High/Scope preschool curriculum comparison study through age 23. *Early Childhood Research Quarterly, 12*, 117-143.

Schweinhart, L. J., & Weikart, D. P. (1998). Why curriculum matters in early childhood education. *Educational Leadership, 55*(6), 57-60.

Slavin, R. E. (1989). Students at risk of school failure: The problem and its dimensions, in: R. E. Slavin, N. L. Karweit & N. A. Madden (Eds.). *Effective programs for students at risk* (pp. 1-20). Boston: Allyn and Bacon.

Slavin, R. E. (2002). The intentional school: Effective elementary education, in: S. Stringfield & D. Land (Eds.), *Educating at-risk students: one hundred-first yearbook of the National Society for the Study of Education, part II* (pp. 111-127). Chicago: National Society for the Study of Education.

Slavin, R. E., & Madden, N. A. (1989). What works for students at risk: A research synthesis. *Educational leadership, 46*(5), 4-13.

Slavin, R. E., & Madden, N. A. (1993). Success for All: prevention and early intervention in elementary schools, in: P. Leseman & L. Eldering (Eds.), *Early intervention and culture. Preparation for literacy. The interface between theory and practice* (pp. 269-284). Paris: UNESCO.

Smeyers, P. (2005). Idle research, futile theory, and the risk for education: Reminders of irony and commitment. *Educational Theory, 55*(2), 165-183.

Stalnaker, R. C. (1999). *Context and content - essays on intentionality in speech and thought*. Oxford: Oxford University Press.

Stringfield, S., & Land, D. (2002). Editor's preface. in: S. Stringfield & D. Land (Eds.), *Educating at-risk students: onehundred-first yearbook of the National Society for the Study of Education* (pp. vii-x). Chicago: National Society of the Study of Education.

Van Goor, R., Heyting, F., & Vreeke, G.-J. (2004). Beyond foundations - signs of a new normativity in philosophy of education. *Educational Theory. 54*(2), 173-192.

Vasiliou, I. (2002). Socrates' reverse irony. *Classical Quarterly, 52*(1), 220-230

Vlastos, G. (1991). *Socrates. Ironist and moral philosopher*. Cambridge: Cambridge University Press.

Winch, C., & Gingell, J. (2004). *Philosophy & educational policy: a critical introduction*. London: Routledge Falmer.

V
EPISTEMOLOGICAL INSIGHTS AND CONSEQUENCES FOR PHILOSOPHY OF EDUCATION II: RELATIVISM, ARBITRARINESS, AND DYNAMIC-DISCURSIVE CONTEXTS

1. Introduction

The results from my endeavor so far have prompted me to further explore a contextualist approach to epistemology, in which justification is embedded in a horizontally ordered network of mutually related beliefs. The first question that springs to mind then is how we can understand the idea of contexts of justification at all. After all, the contextualist authors discussed in chapter two, for instance, have not been able to offer any unequivocal clarity on that matter. The study concerning irony as a philosophical method presented in chapter four might provide an answer for this issue. An exploration of Rorty's interpretation of the concept of irony will help us to obtain an idea of how the embedding of statements in a broader context of related beliefs might be comprehended. The subsequent question to be answered then will be whether the idea of contextual justification does not inevitably lead us into the trap of relativism.

2. Contextualism: relativism or arbitrariness?

In Rorty's interpretation of context claims and arguments are understood in light of a used vocabulary. Here, vocabulary is understood as a linguistic tool that is characterized by a collection of beliefs, rules or principles – with regard to matters like correct use of words, manner of speaking and standards for justification – and is applied in sight of the realization of a certain objective. Each vocabulary has its own forms of justification and description. To give an example: a physicist's vocabulary, which is used, for example, to describe and explain the trajectory of a ball, differs greatly from a competitive athlete's vocabulary. Whereas the athlete's vocabulary will explain the ball's trajectory in terms of the technical treatment by, and the tactical intentions of, the player, those matters are irrelevant to the physicist, who will focus on the forces exerted in and on the ball. Vocabularies, thus, constitute the justification context for beliefs. However, they cannot be regarded as foundations that function as ultimate justification grounds, as in foundationalism. The great difference is that they do not function as justification grounds because they are thought to enable us to describe the world as truthfully as possible, but because they have proven to be of practical use (after all, they are 'tools'). They are, therefore, not regarded as 'certain', but rather as suitable for the task which with we are faced. For the same reason, it is no use to wonder which vocabulary is the 'correct' one, irrespective of what is intended by using that respective vocabulary.

In Rorty's approach, vocabularies seem to be there, independent of, and 'ready-to-use' for a speaker. From that perspective speakers may use a certain vocabulary, but as soon as they do, they also seem to be forced to stay within the boundaries of that vocabulary. Once an athlete's vocabulary is used, the world appears in terms of techniques and tactics, whereas a physicist's vocabulary conjures up a world of exerting forces. This view might seem to imply relativism, which says that things 'are true' only from a certain – cultural or subjective – perspective (in this case, a vocabulary) (cf. Williams 2001, p. 10). The reproach of relativism, however, is out of place here. The concept of relativism demands that certain descriptions and explanations emerge as 'true' when a certain perspective is used. Perspectivity, then, implies the impossibility of any exchange beyond the boundaries of one's own horizon, so that an evaluative comparison of perspectives is impossible and we must conclude that the one perspective - in this case, the one vocabulary - is as good, or as 'true', as the other. In this regard Blackburn emphasizes that the danger of relativism is over as soon as we have been able to explain that an exchange beyond the boundaries of perspectives is indeed possible (cf. Blackburn 2001). And although it might seem that Rorty's vocabularies force perspectives on their users, this is not how he imagines the use of vocabularies. In Rorty's view, a vocabulary is a communicative context. The communication may be restrained by the vocabulary used but, according to Rorty, this does not mean that we cannot look across the boundaries of the vocabulary; for instance, because different vocabularies may be 'played off against each other' (cf. Rorty 1989, p. 73) in light of what we wish to achieve. This also shows that it is not the vocabulary that functions as the final criterion for the 'truth', but rather the extent to which we are capable of dealing with a certain practical problem. Against this background, it is understandable as to why Rorty describes the use of vocabularies as a way of 'coping' (ibid., pp. 14-15). Furthermore, Rorty also deems it possible to create, and further develop, vocabularies. "Languages are made not found", so he argues (ibid., p. 7). Which shows again that he does not think that vocabularies restrict our view in any predetermined way.

Thus, an image emerges of vocabularies that, in principle, are always under the pressure of rejection, or improvement in their function as a tool. In philosophy, such a process of change usually takes place in when there is a "contest between an entrenched vocabulary which has become a nuisance and a half-formed new vocabulary which vaguely promises great things" (ibid., p. 9). In Rorty's view, it is not 'pre-existing' vocabularies that make claims 'true', as would be the case in relativism.

Hence, his approach also leaves no room for the relativist idea of different, coexisting 'truths'. Moreover: Rorty relinquishes the entire traditional notion of 'truth'. Rorty claims that, due to the uncertainty of our fallible vocabularies, it is impossible to achieve something like an 'umbrella-truth', and that the idea of vocabulary-based 'local truths' cannot be upheld, because vocabularies do not set outer boundaries for whatever may, or may not, be considered to be justified. Within Rorty's interpretation of vocabularies, certain descriptions and statements are used not because they are regarded as 'true', but simply because they do their job. And if we are in need of anything better, we just go and look for it.

Rorty, therefore, is not a relativist, but we may wonder whether his approach does not get us into even deeper trouble since he gets us bogged down in a total anarchy of vocabularies where one can just randomly choose to use one or another. We saw earlier that, as regards this issue, Bransen accuses Rorty of 'Spielerei' (cf. chapter four). In his reaction to Bransen, Rorty counters this allegation by drawing attention to the influence of what Gadamer refers to as 'Wirkungsgeschichte' (see Bransen 1991, p. 173). Rorty thereby wants to show that we most certainly can derive criteria from our collective history that enable us to let certain vocabularies prevail over others, or to reject certain vocabularies as undesirable or abject. In other words, he does not see an arbitrary sequential order of vocabularies. He rather sees a history in which vocabularies are continually tried, tested, adjusted and/or replaced as linguistic tools in a continuing communicative process – which he refers to as an ongoing dialogue. In light of human history, but also against the background of communicative exchange in our current society that may be regarded as characterized by plurality and change, this seems to be a plausible picture that shows no signs of 'Spielerei' at all. With Rorty, nothing is certain, but nothing is simply accepted either: everything is potentially exposed to criticism and replacement. That is the central point in Rorty's concept of irony – and also my motivation to elaborate on that concept in chapter four.

Therefore, in order to escape the reproach of relativism and arbitrariness, the potential discussability and changeability of justification contexts is essential. Within Rorty's variant of contextualism, ideas about the practical usability of communicative justification contexts seem to play a primary role in this. After all, the value of the descriptions and statements that we use seems to be evaluated in light of the extent to which they enable us to deal with the practical problems we face. This idea is utterly important because it enables us to abandon the traditional idea that the epistemological value of descriptions and explanations will ultimately depend on the question of whether, or to what extent, they offer us a 'correct' description of the how the world *is*. However, Rorty does not manage to draw a plausible picture of how that process of vocabulary-transformation within the communicative process, which takes such an important place in his vision, might actually work. He does show that he does not regard the formation of a new vocabulary as something common, but rather as something exceptional that requires the creativity of a poet – broadly interpreted as "one that makes things new" (ibid., pp. 12-13), like certain revolutionary writers, scientists or philosopher (cf. Rorty 1989, pp. 19-20). However, his work does not elaborate on the possible origin of this creative impulse, or how it takes shape in the communicative process. Those issues are important because they are where the epistemological significance of the concept of 'context' – including its possibilities and limitations – can only become truly clear. The next question, therefore, is how we can understand the transformation of communicative justification contexts.

In chapter four, drawn upon the ideas of discourse theorist Stalnaker (cf. 1999), I give an explanation of how the transformation of communicative contexts might take shape. To understand this explanation, it is first of all necessary to recognise that a communicative context is not imagined as an independent, ready-to-use entity – which the metaphor of a vocabulary as a tool seems to suggest – but rather as something that is constructed time and time again *within* the communicative process by the participants in that communication. According to Stalnaker, claims and arguments are not understood in light of a clearly defined collection of rules and basic beliefs. Instead, they should be viewed in light of the (assumed) *image* the speaker has of the beliefs (how things are seen, what is considered to be the right tone or a proper argument, what the conversation is about or the direction it should take) his audience shares when he makes his statements. Stalnaker, therefore, defines a communicative context as a collection of what he refers to as 'speaker-presuppositions' (Stalnaker 1999, p. 101). Stalnaker's interpretation of communicative contexts enables us to understand how contexts transform over time. Since communicative contributions inevitably influence the audience - at least, if the audience accepts them as such – the participants in the communication will have to re-imagine the context after each communicative contribution. They will have to gauge how the collection of presuppositions, which the audience is assumed to share, is influenced, because that in turn helps to assess how the next communicative contribution is to be understood. It is in this way that a context does not emerge as relatively independent, since it is constructed by each participant in the communication, and not as unchangeable either, as this (re-)construction will have to be (re-)adjusted time and time again. Moreover, the transformation of communicative contexts does not emerge as something exceptional – as Rorty imagines – but as a structural characteristic of communication and, therefore, as something common. This picture of the everyday (re-) construction of communicative contexts will also be less susceptible to the reproach of relativism, since every suggestion of justification contexts being independent, and determining beforehand whatever is regarded as justified, or not, is completely avoided here.

Brandom (cf. 1994; 2000) develops a position that links up with the theory of Stalnaker. In line with Stalnaker, Brandom emphasizes that the claims and arguments speakers should be understood as attuned to the beliefs that the audience is thought to share, on the one hand, and aimed at changing (some of) those beliefs, on the other hand, so that the communicative contexts will partly transform with *each* contribution to the communication. Since all the participants in the communication will have to keep an eye on how the context is subsequently transformed, Brandom refers to them as 'scorekeepers' (cf. Brandom, 2000). Understood as such, all the participants will thus keep up with the subjects that are discussed, the concepts that are applied, the beliefs that prevail, and the arguments that are deemed to be persuasive within the communication, whilst these issues change from to time to time. Because Brandom uses Rorty's ideas as building blocks, he enables us to expand on Rorty's concept of communicative contexts. This is the reason why, in chapter four, Brandom's ideas play such a decisive role in the formation of my own concept of irony, and eventually in the formation of my own interpretation of context. In becomes clear that Brandom too, not only focuses on the speaker as a user of a vocabulary in his theory, but also, and at the same time, on the audience addressed by that speaker. This puts the spotlight on the ongoing communicative process and the changeability of communicative contexts. Hence, the concept of context that emerges can be regarded as discursive *and* dynamic.

3. A discursive approach to epistemology

The notion of dynamic-discursive contexts of justification changes our image of justification, especially when set off against a foundationalist model of justification. Foundationalists believe that a claim can only find its way to justification if it is supported by a fixed, or 'certain', ground or foundation; an argumentative basis of which the validity itself has already been determined. Williams calls this model of justification a 'prior grounding' model since, according to this approach, one is only "epistemically responsible in believing a given proposition" (2001, p. 24) once its evidence has been established completely. The contextualism that I propose here, submits not only that the idea of an evidential – 'certain' – ground for justification cannot be properly defended, but also that this is unnecessary when it comes to understanding the notion of justification. Claims by speakers are considered against the background of beliefs that assumed for the time being, and that are not claimed to be 'more certain'. On the contrary: most of the time, such assumptions, or presuppositions, remain undiscussed and are not considered at all. That is partly so because speakers are not continually asked to explicitly justify their claims. In communication, it is usually silently assumed that the speakers are justified to make their claims. As William puts it: "[e]ntitlement to one's beliefs is the default position" (ibid., p. 25). It is only when there is reason to in the conversation that someone will be challenged to explicate his or her reasons and those reasons will be scrutinized. Williams, therefore, refers to this as a 'default and challenge' model of justification.

This picture of the process of making and justifying claims resembles the way in which we deal with justification in our day-to-day conversations. Day-to-day conversations are also held against the background of a collection of assumptions that appear to be accepted by all partners in the conversation, and in day-to-day conversation we also usually refrain from asking for explicit reasons for someone's claim. As Elgin puts it: "We just take her at her word" (1996, p. 140). It will even often be viewed as nonsensical, or even ill-mannered if, for no apparent reason, speakers are asked which arguments they believe to support their claims (cf. Adler 2008, p. 345). In such cases, the background of the beliefs and assumptions applied 'by default' functions as a justification ground and thereby provides potential reasons for claims, it is just that we usually do not expect our partners in the conversation to continually explicate those reasons (cf. Elgin 1996, p. 140).

The question now is whether such a concept of day-to-day justification offers also a point of departure for – be it more academic-intellectual – epistemology. Adler thinks it does, because, on the one hand, contributions to conversations can always be deemed to be claims about the world and, therefore, claims to knowledge and, on the other hand, in day-to-day conversations, just like in epistemology, plain rules apply as to when someone is justified to make a certain claim (cf. Adler 2008, p. 337). Adler shows that a so-called 'conversationalist epistemology' is definitely not without requirements, even though it abandons the requirement of an extensive justification of knowledge claims irrespective of the specific communicative context. In his view, in day-to-day conversations strict requirements are set in general to whatever someone may or may not claim. "Hearers would not accept a speaker's assertion if it was doubtful that the speaker has sufficient reason or knowledge to believe his assertion" (ibid., p. 345). As such, it applies to every conversation that the participants involved ensure that not everything is just taken for granted. Both in day-to-day and more intellectual conversations, participants will keep an eye on whether they trust the speakers to have sufficient reasons to make their claims. Of course, whatever is considered to be 'sufficient reasons' in a day-to-day conversation will substantially differ from those in an academic-intellectual conversation, but the justification *model* used does not necessarily have to be different.

Therefore, there is reason to take the 'default and challenge' model of justification connected to my dynamic-discursive concept of context as a point of departure for my own epistemological position. Since this epistemological position depends on a description of the way in which justification takes shape in communicative processes in which claims are made, challenged, and defended, it may be called 'discursive'. One of the characteristic starting points for such a discursive epistemological approach is that the development of knowledge is made fully dependent on a cooperative, critical-reflective exchange between speakers and their audience. The exchange may be regarded as cooperative, since the audience usually trusts a speaker to have sufficient reasons to make their claims, and because speakers aim to meet the expectations of their audiences (cf. Adler 2008, p. 339). The exchange is critical-reflective, for any contributions that are not considered to meet the epistemological standards applied, will be critically challenged. In a case like that, a speaker will be obliged - at least, if they want to be taken seriously - to respond in accordance with the acceptability standards prevailing within the communication (cf. Williams 2001, p. 25). However, critical contributions to the communication do not have a privileged status (cf. Adler 2008, p. 345). The raising of critical questions is simply a contribution to the communication as any other and is thereby also bound by the standards employed in the communicative process.

In a discursive epistemology, the exchange and justification of knowledge is thus made dependent on a community of discourse participants who jointly aim to keep the communicative process going, but that at the same time acts as a gatekeeper that watches closely as to whether the prevailing epistemological standards are met.

The subsequent question is whether whatever is regarded as knowledge, then, is not made fully dependent on an arbitrarily agreed convention between the participants within a communicative community and, if so, whether we can still speak of a progress of knowledge in a more scientific sense. On that issue, chapter six, which revolves around the question of what it means for a subject to acquire a language, will offer us insights that can help us along.

4. Provisional remarks about the tasks of the philosophy of education

Against the background of the central question of this thesis, it is now relevant to check whether conclusions may already be drawn as regards the significance of the epistemological insights found for my ideas about the tasks and possibilities of a future philosophy of education.

We have arrived at an approach in which epistemological justification is made dependent on dynamic-discursive communicative contexts. It, therefore, seems that in any case the ambition to generate generally valid claims should be abandoned. This conclusion also affects the discussion about whether or not general prescriptive pretensions can be connected to philosophical educational claims, especially since such pretensions formerly were regarded as depending on epistemological validity. On this issue, it has in any case become clear that there is no longer an epistemological ground for the formulation of general educational prescripts. In as far as philosophers of education feel the need to make general educational recommendations, they will have to appeal to considerations other than epistemological ones – although it is difficult to see how one could do this in a philosophically acceptable way.

The antifoundationalist authors in chapter two have already drawn attention to the corrosion of the prescriptive possibilities of the philosophy of education. They plead for a philosophy of education that is not aimed at the substantiation of educational claims in order to offer these as recommenddable, but rather at the clarification of the restrictions to which every possible form of substantiation, of justification is bound, and so to fight the systematic exclusion that would be implied by such a justification. Although this thesis, partly due to the epistemological insights of these antifoundationalist philosophers of education, has put me on the track of developing my own contextual approach to knowledge, contrary to these authors I do not see a direct reason to attach any content-specific consequences to it. I have submitted that communicative contexts continually transform with every communicative contribution, so the plea for the clarification of contextual restrictions as a kind of exclusive possibility to break through the boundaries of justification contexts seems redundant. Insofar as the philosophy of education wants to engage in the clarification of elements of the justification context – for instance by applying an ironic philosophy – it is also bound by a communicative context, which implies that it will not help us in overcoming justificatory restrictions anyway.

Now, does this mean that the entire epistemological exercise is irrelevant to the question of the tasks of the philosophy of education? In response to the ideas of antifoundationalism in the philosophy of education, Cooper claims something to that effect. Insofar as they would be valid, he argues, these ideas are of such a nature that when it comes to the interpretation by philosophers of education, "[they] leave everything as it is" (Cooper 2003, p. 212). I would not want to go that far. Considering the foregoing, for instance, it seems that it is quite well-defendable to act with some caution where your prescriptive pretensions as a philosopher of education are concerned. Whether there is more to say about the content of the philosophy of education in the light of my epistemological insights, I will discuss later in this thesis.

5. References

Adler, J. E. (2008). Conversation is the folks' epistemology, *The philosophical forum, 39*, 337-348.

Blackburn, S. (2001). *Being good: A short introduction to ethics*. Oxford: Oxford University Press.

Brandom, R. B. (1994). *Making it Explicit: reasoning, representing, and discursive commitment*. Cambridge (Mass.): Harvard University Press.

Brandom, R. B. (2000). *Articulating reasons: an introduction to inferentialism*. Cambridge, MA: Harvard University Press.

Elgin, C. Z. (1996). *Considered judgment*. Princeton: Princeton University Press.

Rorty, R. (1989). *Contingency, irony, and solidarity*. New York: Cambridge University Press.

Stalnaker, R. (1999). *Context and content. Essays on intentionality in speech and thought*. Oxford University Press: New York.

Williams, M. (2001). *Problems of knowledge*. New York: Oxford University Press

VI
NEGOTIATING THE WORLD. SOME PHILOSOPHICAL CONSIDERATIONS ON DEALING WITH DIFFERENTIAL ACADEMIC LANGUAGE PROFICIENCY IN SCHOOLS[8]

1. Introduction

Partly because of globalization and accompanying processes of migration, educational systems are confronted with the problem of differential academic language proficiency of children within school classes. This problem is an issue of major educational concern these days. Discussants not only consider it disturbing it interferes with pupil performance, (Bankston & Zhou, 1995; Calderón, Hertz-Lazarowitz & Slavin, 1998; Eisenstein Ebsworth, 2002; Garcia, 2002; Jepson Green, 1997; Porter, 2001; Reyes & Rorrer, 2001; Thompson, 2004; Valdés, 2002; Wright, 2004), but also because it is related to problems of social prospects (Bankston & Zhou, 1995; Porter, 2001; Thompson, 2004; Valdés, 2002; Wright, 2004), and citizenship (Thompson, 2004). This cluster of problems that is related to differential academic language proficiency raises a lot of questions, including philosophical ones. Such questions may concern relations between social groups and how to assess them, the position of the individual in society, or equality of opportunity, to mention only a few examples. Philosophical questions with respect to differential language proficiency may also concern the definition of underlying concepts, that are used in formulating the problem as such, e.g. the concept of language or that of lemming. Though I realize, that both categories of philosophical questions will turn out to be interrelated, my initial interest concerns the second category. In particular, I want to investigate what implications our conceptions of learning a language have for the definition of the language-learning subject. In other words, I am interested in the definition of the human subject as related to the learning and use of a language. To that end, I will develop a conception of language acquisition, and formulate its implications for the definition of the subject as I go along. In addition, I will discuss the consequences of my results for understanding differential linguistic development of children in schools, and for dealing with the problem at the level of policy.

Starting from Ludwig Wittgenstein's 'meaning-as-use' theory, in

---

8 Published as: Van Goor, R. & Heyting, F. (2008). Negotiating the world. Some philosophical considerations on dealing with differential academic language proficiency in schools. *Educational Philosophy and Theory, 40*(5), 652-665 (printed with permission)

the next section I first introduce a conception of language acquisition as initiation in a language community. In my analysis learning a language appears as a process, in which those who undergo the process of initiation essentially play an active and also potentially renewing part. This brings us to the relevance of the experience the language-learning subject has of the 'world' - of that which is spoken about - for becoming an active participant in a language community, which I investigate in a third section. Here I find that the development of judgmental skills with respect to 'the world' constitutes an important ingredient of language-learning as well. Both dimensions of language-development - participating in a language community and developing judgmental skills - seem closely interrelated. The interdependencies of the three components that come into play in language-acquisition - subject, language-community, and world - are discussed in a following section. Here, the language-acquiring subject will appear as an intentional participant in a socially mediated process of negotiation about what will be accepted as the world. From this process, the world emerges as a conceptualized and a communicable complex of meanings, in ever temporary forms that keep being renewed in the process of communication time and again. In a concluding section, I discuss some consequences of my conclusions for the educational meaning and relevance of differential academic language proficiency, which also throws some light on current differences of opinion about which policy to take in this issue.

2. Language acquisition as initiation

Developing an idea of the language-acquiring subject requires a conception of language to begin with. Wittgenstein's meaning-as-use theory offers a good candidate that is hardly controversial in philosophy of education. Developing an understanding of the analysis of concepts Richard Peters (1966) first drew attention to this theory, and it seems hardly possible to imagine philosophy of education without it ever since. Based on a use-theory of meaning Peters also developed an understanding of initiation into language-use, however, in this case without explicit reference to Wittgenstein. More recently philosophers of education use Wittgenstein's theory in particular when dealing with problems of initiation into meaning and the use of language (see, for example, Hamlyn, 1989; Marshall, 1985; Smeyers, 1995; Smeyers & Marshall, 1995; Spiecker, 1977).

The idea of the later Wittgenstein, that we should not understand words and their meanings as direct representations of the world of objects 'out there', excludes any conception of language-development that is based on the idea of learning to apply the right word to the right object.

Wittgenstein's (later) philosophy of language recognises a great variety of language uses (cf. Wittgenstein, 2001, § 23). In descriptive language use, reference to the world is not considered a necessary precondition for meaningful speech about the world. One of Wittgenstein's examples to explain this addresses the question how we learn words like 'pain', i.e. words that can seemingly only draw from one's own experience. Wittgenstein points out that this only is a problem as long as we use the model of 'object and designation' to understand the learning of such words. However, on closer inspection the object appears to drop out of consideration as irrelevant (Wittgenstein, 2001, § 293). Wittgenstein illustrates this with the use of the word 'beetle'. As one can only access their own experiences, Wittgenstein explains, one can at best find out in what situations - *as experienced by the subject* - other people seem to use the word 'beetle'. Wittgenstein: "Suppose everyone had a box with something in it: we call it a 'beetle'. No one can look into anyone else's box, and everyone says he knows what a beetle is only by looking at *his* beetle. - Here it would be quite possible for everyone to have something different in his box. One might even imagine such a thing constantly changing. - ... That is to say: if we construe the grammar of the expression of sensation on the model of 'object and designation' the object drops out of consideration as irrelevant" (Wittgenstein, 2001, §293). In the end, we do not learn the meaning of the word 'beetle' by being confronted with the object, but by being confronted with the use of the word 'beetle'.

Learning linguistic meaning seems primarily an issue of getting familiar with the conventions regulating linguistic usage. Consequently, language acquisition can be understood as a process of initiation into the customs of a linguistic community. However, this process would be misunderstood if compared to a simple transfer of the rules governing customary usage, because these rules are not such that mechanical application would be possible, or that they would guarantee correct linguistic usage. "For not only do we not think of the rules of usage ... while using language, but when we are asked to give such rules, in most cases we aren't able to do so" (Wittgenstein, 1969a, p. 25). In addition, Wittgenstein says: "The man who is philosophically puzzled to see a law in the way a word is used, and trying to apply this law consistently, comes up against cases where it leads to paradoxical results" (Wittgenstein, 1969a, p. 27; see also Wittgenstein, 2001, §198 ff.). Linguistic customs are changeable and not unambiguous at that. Wittgenstein's concept of 'family resemblances' expresses this as well. As a consequence, in learning a language the subject is not learning to more or less passively reproduce specific customs. As Jerome Bruner (1990), formulates it: "Language is acquired not in the role of spectator but through use. Being 'exposed' to a flow of language is not nearly so important as using it in the midst of 'doing'" (p. 70). Language-acquisition requires 'doing', active participation.

Publications about differential language proficiency in schools demonstrate this idea of learning by doing as a widely accepted starting point. For example, Eugene Garcia says. that 'natural communication situations must be provided' (Garcia, 2002, p. 23), whereas Elise Jepson Green thinks it important to create a "language rich environment ...with good models of English language use" (Jepson Green, 1997, p. 152), and Pedro Reyes en Andrea Rorrer recommend offering opportunities for "participation in meaningful interaction" (Reyes & Rorrer, 2001, p. 169). These authors share the assumption, that language acquisition can best be furthered directly related to actual communicative contexts. A transfer-model of language acquisition only seems implied where discussants recommend separate instruction, for example to explain grammatical aspects of language (Eisenstein Ebsworth, 2002, pp. 107-09). Similarly, in educational policy discussions some Dutch discussants (Vaessen, Hoogeveen & Stassen, 2003) argue in favor of separate classes for pupils with limited academic language proficiency in order to improve the preconditions for their school performance. Of course, such classes could make use of the principle of 'learning by doing' as well, but the idea of first improving language proficiency before concentrating on further academic achievement still separates the context of language acquisition from the context of language use.

In order to find out what this active participatory role of the subject in acquiring a language implies for our conception of the language-learning subject, we will need a sharper view of what this activity entails. In his interpretation of Wittgenstein's contribution to this issue, Paul Smeyers (1995) distinguishes two dimensions in this active participation. The first dimension follows logically from the idea, that usage does not consist in following fixed and unambiguous rules. As a consequence, the novice is faced with the task to develop his own pattern of meaning from the more or less diffuse instances of usage he encounters in a variety of contexts. As the subject is not confronted with linguistic customs as such but only with scattered instances of usage, he will have to reconstruct some version of these customs for himself from the cumulative series of linguistic experiences he is confronted with. This dimension of active subjective participation in acquiring a language results in a unique connotation, a 'personal color" (Smeyers, 1995, p. 123), for every new word or expression an individual gets familiar with.

The second dimension Smeyers distinguishes in the active participation of the language-acquiring subject is related to whatever language is about, to the object or situation language users attach a meaning to. To explain this, Smeyers refers to an example of Wittgenstein, discussing the question how anybody could recognise the taste of sugar as the taste of 'sugar' (Wittgenstein, 1980, §353). This also requires an activity of the subject, if only by making an appeal to his memory where the taste of sugar would be stored for comparison with future experiences. However, in reality recognition will never be as simple as that, as Smeyers emphasizes, because situations will never be completely identical - even if restricted to tasting sugar. Consequently, recognizing something as 'sugar' will not only require an appeal to memory; the subject also has to construct the situation as an instance of tasting sugar, or, as Smeyers (1995, p. 123) expresses it, he has to accomplish a "performing of the meaning" as well. The object the use of language is about is never given as such, but only results from this 'performance of meaning'.

After the preceding, language acquisition can be understood as a process of initiation in the customs of usage by way of active participation. The active contribution of the novice to this process includes the development of complexes of meaning, characterized by unique personal coloring, which will also contribute to the renewal of linguistic customs in the language community concerned (Smeyers, 1995, p. 123). In addition the novice actively contributes to the process by relating his own sensetions of the 'world' to these developing complexes of meaning and language and vice versa.

The first activity of the language-learning subject - reconstructing possible meanings of linguistic expressions - primarily relates the learning subject to the present community of language users. The second activity - relating sensations to linguistic expressions - primarily relates the learning subject to the world, to the objects and situations language is about.

Though I explained above, that Wittgenstein's theory of meaning does not presuppose reference to the world as a necessary precondition for meaningfully speaking about it, a theory of the language using and acquiring subject seems to require an explanation of the relation between the subject and the world of objects in order to understand "the grammar of the expression of sensation" (Wittgenstein, 2001, §293). I will explore this relation between subject and world in the next section; after that I will be able to give a more detailed description of the way the language-learning subject is related to other language users.

3. Language acquisition as development of judgment

Though learning to use a language requires a relationship to a community of language users, paying attention to that which is given meaning to in using language, is just as indispensable for understanding the use of language. In order to develop a conception of the language- learning subject, not only his developing ability to make use of linguistic devices is important, but also the experience - in the broad sense of the word - that is expressed while using language. In explaining his Wittgenstein interpretation, Michael Luntley elaborates on the role of this 'experience with the world' in linguistic development. Distinguishing this experience from the influence of other language users, he says: "Others help it phrase its engagement, but they are not the object of engagement except in so far as others are, of course, part of the world" (Luntley, 2003, p. 172). This developing 'engagement with the world' is an indispensable part of linguistic development that will also affect the role of the language community in the process. According to Luntley's interpretation of Wittgenstein, the use of language is "calibrated in patterns of use that manifest a grip on that which is independent of will" (Luntley, 2003, p. 135). Consequently, the wish to get a grip on 'that which is independent of will' constitutes an important source of any linguistic development.

In this view, the use of language implies an intentional turning of the subject towards the world. In this respect the language-using subject plays an active part. Luntley points out, that according to Wittgenstein's view of language, the subject can never have a direct and unmediated connection to the world. It is an attitude of the subject, crystallized into an 'attentional awareness' as Luntley calls it, which shapes the connection of the subject with the world. In other words, the way the subject focuses on the world determines the resulting 'coupling' between subject and world. According to Luntley, an essential developmental process with respect to language acquisition takes place at the level of these 'couplings'. In particular, this developmental process concerns the development of 'attentional skills', which will give rise to more sophisticated forms of attentional awareness, which in its turn will cause a developmental change in the couplings between subject and world. According to Luntley, these developing attentional skills are of vital importance to the linguistic development of a subject.

An example of the coupling between subject and world as based on the use of attentional skills can be extracted from John White (2002, pp. 32 ff.). Like Luntley, he relates language acquisition to a process of mediated adjustment to the world. In White's formulation, this process starts at a very early stage, based on pre-linguistic judgmental processes he calls 'sign recognition', which can already be observed in babies who seem to recognise the breast presumably as a sign for food (White, 2002, p. 40). Luntley (2003; 2004) refers to this kind of recognizing ability that is fundamental to any descriptive conceptual development, as the ability of 'seeing similarities'. In his view, "the notion of seeing the similarities is

106

primitive for Wittgenstein" (Luntley, 2002, p. 271; see also Luntley, 2003, p. 77f.). The refinement of this ability of 'seeing similarities', as a part of developing attentional awareness, is important to the conceptual and linguistic development of the subject, according to Luntley. For example, in practical activities like sports, an advanced practitioner will be able to distinguish and recognise a much more detailed pattern of relevant signs than a beginner (White, 2002, p. 42), and this ability will also support the refinement and nuance of the use of linguistic means. Against this background, it becomes understandable why Luntley describes conceptual development as a training in "seeing things aright" (Luntley, 2003, p. 173) or as an "apprenticeship in judging" (Luntley, 2004, p. 7). He has a process in mind of learning to see ever more differentiated similarities, on the basis of a preceding attitude - being hungry, wanting to hit the ball - of the subject.

In this way, Luntley relates conceptual development (or at least development with respect to 'demonstrative concepts', cf. Luntley, 2004, p. 10) primarily to the development of those "attentional skills by which we see things as similar" (Luntley, 2004, p. 7). In his view, this also affects the way we should understand the role of the community in acquiring language proficiency. According to Luntley, this community is primarily relevant with respect to the development of 'attentional skills', whereas he downplays the role of the community as provider of a set of rules for the correct use of language. "For such concepts, possession of the concept ... does not consist in a possession of a theory that drives the seeing. Hence, the model of conceptual development for such concepts has to be construed as an apprenticeship in seeing" (Luntley, 2004, p. 7). The process of 'joint attention' constitutes the core mechanism by which the community can further the development of capacities to see similarities. According to Luntley, this process results from the "fundamental impulse to share" (Luntley, 2004, p. 5), and it is directly related to the intentional relation of the subject to the world. This impulse to share constitutes the primary motive for linguistic development (Luntley, 2004, p. 172).

In as far as limited academic language proficiency is caused by inadequately developed attentional skills, Luntley's approach implies a specific explanation for the connection between poor pupil performance and limited language proficiency. From this perspective, learning problems would not so much result from a limited familiarity with the conceptual tools that are used in the classroom, but rather from limited attentional skills with respect to those realms of reality school education refers to. Many discussants in the field mention this kind of explanation for the inhibiting effects of limited academic language proficiency

(Bankston & Zhou, 1995; Calderan et al. 1998; Eisenstein Ebsworth, 2002; Garcia, 2002; Jepson Green, 1997; Porter, 2001; Reyes & Rorrer, 2001; Thompson, 2004; Valdés, 2002; Wright, 2004). For example, Garcia seems to imply this view in considering linguistic development as expressing a capacity "for functioning in and perceiving the world" (Garcia, 2002, p. 27). The way he understands the importance of preserving each individuals language and culture of origin also seems in accordance with a Luntleyan perspective. From that perspective, the cultural capital a native language incorporates should be understood as a collection of attentional skills, from which the development of (a second) academic language would also tap. This approach can be heard in Garcia's proposition, that "many of the strategies that children use to acquire this language seem to be the same as those used in first- language acquisition" (Garcia, 2002, p. 21).

Not only the relevance of attentional skills for developing language proficiency, as emphasized by Luntley, has consequences for education. His view of the 'fundamental impulse to share' - expressed in joint attention processes - as a primary motive for linguistic development, does not remain without consequences either. In order to establish effective 'joint attention', adapted to the attentional awareness of the pupil, stimulating development in the linguistic domain will make high demands on the quality of communicative relationships between teacher and novice. As Luntley formulates it: "One aspect of this way of looking at things might be to suggest that teachers need to move closer to the communicative styles of parents with their own children, since instructtional conversations will require highly refined interpersonal competencies in scaffolding the attentional frames of different individuals" (Luntley, 2004, P. 17-8). Applied to policy with respect to differential academic language proficiency, this means that - apart from assuring active participation in a language community - attuning to the direct engagement with the world of the developing subject should be secured. This aspect can be found in Garcia as well. If the school does not succeed in connecting with the "representations of the child's world it is negating the tools" that are a precondition for the linguistic development of the child (cf. Garcia, 2002, p. 27).

The introduction of the engagement with the world in the model of developing language proficiency also affects the position of the language community in the process. To put it shortly, it shifts attention from getting familiar with the conventions of usage to the more fundamental attempt to get a grip on the world as "that which is independent of will" (Luntley, 2003, p. 135), and to the attentional skills that are required to that end. From this perspective, furthering linguistic development starts with scaffolding the process of 'coupling', of the "relation between the subject's egocentric take on things and how things are" (Luntley, 2004, p. 10-11). This way of subordinating linguistic development to enga-gement with the world seems consistent with a Wittgensteinian view of language as a 'tool'.

This necessary engagement of the language learning subject with the world will bear upon the issue at hand, how to deal with differential academic language proficiency in schools. If developing attentional skills should be understood as tailoring the subject's take on things, Luntley's argumentations implies a plea in favor of a child-centered approach to the child's engagement with the world would be a didactical instrument for realizing the same aim in any child: getting it better geared to this world. In order to better understand the role of the subject in its linguistic development, I will have closer look at the way this process of concept-tualizing the world takes place.

4. Linguistic development as participative negotiation
Wanting to help children develop a sharper view of things, is a familiar way of approaching education. From this perspective, a mother will draw the attention of her baby to an object in the room, or a trainer will alert to as yet unnoticed details in the flight of a ball, both expecting a contribu-tion to improved attentional skills from such actions. However, in his *On Certainty* Wittgenstein (1969b) warns of the tendency to adopt the idea that words are representations of objects in an external reality - as it implicitly functions in everyday usage - unquestioningly as a basis for philosophy as well. His example of a child that is learning to use the word 'tree' can explain this. 'Clearly no doubt as to the tree's existence comes into the language-game', Wittgenstein (1969b, §480) says. However, he adds, as soon as a philosopher - here personified by Moore - would say "'I know that that's a tree' one suddenly understands those who think that has by no means been settled" (Wittgenstein, 1969b, §481). "It is as if 'I know' did not tolerate a metaphysical emphasis" (Wittgenstein, 1969b, §482).

At this point, I should remind of the fact, that Luntley - in accor-dance with Wittgenstein - does not suggest the possibility of an unme-diated grip of reality either. In addition, he does reject a purely conven-tionalist interpretation of Wittgenstein's view of language, that reduces conditions for correct usage to the language community alone (Luntley, 2003, p. 17), because that would render the idea of aboutness' incompre-hensible. He can only explain this 'aboutness', that is characteristic of any use of language, from the role of the subject as an active agent, taking an attitude to make best sense of his ongoing confrontation with things, an active attitude towards the impediments that go out from the world upon our behavior (Luntley, 2003, p. 2).

The question is, now, how to imagine those 'impediments that go out from the world', presupposing Wittgenstein's view, that referring to the world is not a necessary precondition for meaningfully speaking about it? At this point the way Anthony Rudd refers to a passage in *On Certainty*, where Wittgenstein speaks about practicing mathematics, may be of help. He says: "What it might mean to say that numbers exist is given within the practice of mathematics; there is no sense to the idea that mathematics needs to be validated by a further philosophical proof that numbers do actually exist. ... There is no further intelligible question as to whether those numbers really exist or not. The metaphysical belief in the existence of numbers gives us only a vague and misleading picture" (Rudd, 2003, p. 84-5). In Rudd's interpretation, Wittgenstein would put *practices*, including their implied presuppositions about the world, in the position Luntley' seems inclined - however cautious - to assign to the world itself. Though the example is about mathematics, other practices - including those involving sensory experiences, perceptions, or moral and aesthetic experiences - are not principally different in this respect. In all cases, reference to the unconceptualized world is not necessary.

This concentration on practices, on practical contexts of language use, does not cause the problems Luntley sees in a purely conventionalist approach of the use of language, in particular that it would reduce constraints on correct usage to the whims of an accidental language community. Like Rudd, Nelson Goodman emphasizes the impossibility of drawing a line between language-dependent and language-independent world features. To explain this, he confronts his critic Isreal Scheffler - who questioned his conception of world-making - with the rhetorical question to indicate "which features of the stars we did not make" and "how these differ from features clearly dependent on discourse" (Goodman, 1996, p. 145). According to Goodman this does not imply that "whatever we make we can make any way we like" (Goodman, 1996, p. 145). He agrees, that we will always be constricted by the available material, but he does not localise this material in the world 'out there', but in residues of established ways of imagining it instead, in "scrap material recycled from old and stubborn worlds" (Goodman, 1996, p. 145). What I - following Wittgenstein - called 'practices' appears in Goodman's termi-nology as processes of making and remaking essentially conceptualized 'worlds' (in plural). In this image, constraints result from those aspects of a 'world' we leave intact, thus excluding them from the process of remaking - for the time being. To put it differently, constraints for correct usage are not understood as ontological constraints, but rather as pragmatic constraints - as in Wittgenstein's conception of language as a tool. Accordingly, the "impediments that go out from the world" refer to these practice-related constraints.

Relating linguistic usage to the contexts of specific practices, makes engagement with the world a primarily concept-driven process (Brandom, 2000, p. 26), and not a world-driven process, as Luntley sometimes seems to suggest. Linguistic development remains related to 'seeing things right', as Luntley calls it, but the term 'right' refers to the world as it appears in the practice at hand, not to the world 'as it is'. A further implication of this pragmatic view of practices, including their presuppositions about the world is, that those presuppositions will no longer be considered completely beyond the reach of language users. According to Williams (2004, p. 95), these themes recur in Wittgenstein's refutation of idealism. The use of language within a practice does not only consist in elaborating upon accepted presuppositions about the world, and in testing claims against them. It can also result in bringing to light some of those - mostly implicit - presuppositions, hitherto functioning as constraints to the acceptability of our descriptive claims, but now being subjected to a process of doubt and reformulation. From this perspective, linguistic development, as active world-oriented participation in a language community, becomes the connotation of practically engaged participation in processes of 'world-making'. Wittgenstein's example of the riverbed as the constraints of language-use, and the river as actual usage, is a nice illustration of this, because actual usage is not only restricted by the riverbed - accepted presuppositions about the world - it also affects its course. In addition, the "riverbed of thoughts may shift": what functions as the riverbed at one moment, can become fluid again and become disputable (cf. Wittgenstein, 1969b, §§ 94-99).

This view of linguistic development as growing to full participation in processes of negotiating the 'world' is illustrated by Carol Fleischer-Feldman (1987). Inspired by the philosophy of Wittgenstein and Goodman, she describes this process as it takes place in young children. To that end, she analyses transcribed conversations with young children, demonstrating that their participation in the process is not restricted to considering, commenting, doubting and/or elaborating upon the contributions of their adult partners. They also give evidence of being actively engaged with the implicit presuppositions behind the conversation - the constraints on acceptability - in a twofold way. First, contributions of the children to the conversation give evidence of what Fleischer-Feldman (1987. p. 136) calls 'ontic dumping'. In this process, they make use of earlier contributions from their partners in order to formulate their own subsequent contributions - apparently incorporating those earlier contributions in their collection of ready-for-use presup-positions. In the context of education. this is a rather familiar process. For example, a pupil will first be invited to get familiar with the Pythagorean theorem, on the assumption that he will subsequently be able to use it as a self-evident instrument for solving new problems. The second way in which children give evidence of being actively engaged with the implicit presuppositions of a practical conversation consists in a reverse procedure, Fleischer-Feldman (1987, p. 138) calls 'dump topicalization'. In this process, children subject former - ontic dumps' to renewed procedures of doubt and discussion.

According to Fleischer-Feldman, linguistic development is related to this duality of what is presumed as being the world - the ontic - and what is being discussed on the basis of that - the epistemic. "Ontic structure and status is thus always a stipulation, by which I mean that it is one of many possible ways of construing a situation that is taken as given for the epistemic purposes at hand" (Fleischer-Feldman, 1987, p. 135-6). In this process, each of the participants, including novices, is actively engaged in constructing and reconstructing the presuppositional basis of the practice at hand. Of course it remains possible to distinguish the role of the 'expert' from that of the 'pupil', but not as simply and unambiguous as Luntley's theory would suggest by assessing the quality of attentional skills in terms of the world 'as it is'. In my revised approach, 'expertise' rather refers to the degree of familiarity with the presuppositions about the world of a specific practice. One should be aware, however, that these presuppositions are not unassailable, but can be revised, even through the agency of a novice.

The changeability of presuppositions about the world also affects the concept of 'expertise' in a second way. This is mentioned by Jan Bransen (2002), discussing Robert Brandom's theory of language. Whereas a child - or a novice in a specific field - may primarily be seeking to refine his familiarity with the presuppositions of a practice in order to attain full participation in the game of formulating and discussing claims (in Brandom's terminology: 'giving and asking for reasons'), the expert, as a full-blown participant, could be more inclined to 'dump topicalisation', to call implicit presuppositions of the current discourse - the props of accepted claims - into doubt, deriving satisfaction from questioning "some of the entitlements he used to count on his score" (Bransen, 2002, p. 389). In that sense an expert may be more uncertain than the novice, and his expertise could consist in stimulating doubts instead of certainty. In as far as linguistic development involves a process of adjustment to the world as it is supposed to be - a 'world' that can be subjected to doubt and revision at any moment whenever the current practice, the game of giving and asking for reasons, gives rise to it. Though the expert may be more equipped for it and more inclined to it, experts as well as novices can initiate such spells of revision. Margarita Calderón, Rachel Hertz-Lazarowitz, and Robert Slavin's views of dealing with the problem of differential language proficiency seems well in tune with this philosophical view of linguistic development. They say: "for students to reach high levels of proficiency, they must engage in a great deal of oral interaction, jointly negotiating meaning and solving problems" (Calderón et al., 1998, p. 154). The concept of 'negotiation' emphasizes a specific characteristic of the relation between participants in a language community, including novices: it emphasizes how the 'world' is not an unassailable basis, but an always provisional product of the communication process itself.

5. Conclusions and comments

By way of conclusion I summarise what linguistic development requires from the developing subject. Apart from developing a personal pattern of meanings from the linguistic instruments a language community has on offer, and from applying them to new experiences, the linguistically developing subject also has to accomplish the identification of his experience as an instance of a specific 'something'. Luntley explains this as a kind of adjustment to the world, supported by attentional skills, that can be improved through procedures of joint attention. Because of the impossibility to distinguish the world 'as it is' from the 'world' we presuppose it is, this identification of experience should not primarily be understood as a result of 'seeing', but rather as a primarily concept-driven process. Following Brandom, I am inclined not to consider 'seeing similarities' - perceptual judgment - as the primary basis of linguistic development, but a kind of concept-based judgment instead. Sharply distinguishing the use of language from anything like a pure reaction to the world, Brandom denies phenomena of simple sign recognition the status of (primitive) concept-possession. Whereas one could attribute the ability of sign recognition to a thermostat, it would make no sense to consider a thermostat as a device that is in the possession of concepts (Brandom, 2000, p. 48). In order to react, the thermostat does not need a concept of temperature, because - unlike language users - it does not have to identify its experience of the situation in terms of temperature. Therefore, the thermostat lacks an ability, any concept-user will need.

Linguistic development requires that the subject develops an ability of understanding, conceptual judgment with respect to the question as 'what' a situation should be identified. This makes linguistic development principally embedded in a practice and its implied presuppositions about the world. The development of this capacity for conceptual judgment is related to the language community and the conceptual tools for describing the world that are developed in its various practices. This makes a language community not only a provider of linguistic devices for ends of communication, but also a workshop of conceptualized worlds, that will serve as bases for assessing claims. As a consequence, the relation of a subject and a language community is not only an issue of one-way traffic, in which the subject gets familiar with the customs of this community. The relation between subject and language community also has a character of practice-related negotiation about the definition of the world in the context of a specific practice. Against this background, the image of a 'language-game' can be done full justice, because we can now take into account, that speakers are not only engaged in a process of reporting, but at the same time in a process of defining and influencing what presuppositions about the world participants will share.

This approach of linguistic development has consequences for dealing with differential academic language proficiency, that deviate from usual practice. As mentioned before, most authors suggest an approach involving some kind of immersion in the practice of academic language. Though the results of my investigation do not deny this approach, they would lay more emphasis on bringing children in practical situations, situations of cooperative problem-solving, requiring active participation in language games, in processes of developing descriptions of the world and revising the criteria for doing so, including discussing the definition of the problem itself. This last dimension particularly seems to be lacking in educational discussions about academic language proficiency. Though some authors seem to recognise the relevance of active participation, the definition of the situation itself invariably seems to be considered unquestionable. In such cases, 'negotiation' - as Calderón et al (1998) call it - would mainly refer to the idea, that the descriptions of the world in a language community are always a provisional product of the communication process itself In their view, being involved in this process would best stimulate linguistic development. For example, Calderón, Hertz-Lazarowitz, and Slavin recommend a didactic approach, that provides opportunities for students to work together in discussing practical problems and ways of formulating them (Calderón et al., 1998, p. 154) because of its positive effects on developing academic language proficiency.

Limited academic language proficiency appears not only related to lacking familiarity with the conventions of school language, though some authors almost exclusively pay attention to this dimension, as in Rosalie Porter, who says with respect to language proficiency: "It is sound educational policy to require one objective, uniform measure of student achievement as a prerequisite for high school graduation" (Porter, 2001, p. 409). Academic language proficiency seems also related to the degree of familiarity with prevailing presuppositions about the world in the context of this particular practice, the school. Realizing that these prevailing presuppositions result from an ongoing communicative process would put novices in a different position - less assimilatory in nature, while recognizing their active participation at a more fundamental level. Linda Thompson (2004) seems to recognise this, where she pleads in favor of a procedure that would make the organization of school education a subject of discussion between representatives of different language communities. However, her recommendations make a halt before the practice of teaching itself even starts, which excludes the process of linguistic participation from the actual teaching practice again.

Of course, my suggested theoretical model in which academic language is considered instrument as well as product of communicative practices, does not exclude the possibility of considering assimilation of linguistic minority groups a desirable approach - if only to protect traditional definitions of the world. Despite of that, the desirability of such an approach seems questionable from my point of view. It could be important to all participants - novices and proficient pupils - to be recognized as participants in the processes of communication that are involved in school education. This view suggests to regard participation in linguistic practices not merely as an instrument for the improvement of the academic achievement of linguistic minority groups, but to regard it an educational end in itself for all pupils. Becoming confronted with situations of doubt and (re)description of the world- for instance triggered by differences in language and culture - will leave all pupils better prepared for actively participating as voting citizens in contemporary globalizing societies.

6. References

Bankston III, C. L. & Zhou, M. (1995). Effects of minority-language literacy on the academic achievement of Vietnamese youths in New Orleans. *Sociology of Education, 68*, 1-17.

Brandom, R. B. (2000). *Articulating reasons. An introduction to inferentialism*. Cambridge: Harvard University Press.

Bransen, J. (2002). Normativity as the key to objectivity: An exploration of Robert Brandom's Articulating reasons, *Inquiry, 45*, 373-392.

Bruner, J. (1990). *Acts of meaning*. Cambridge: Harvard University Press.

Calderón, M., Hertz-Lazarowitz, R. & Slavin, R. (1998). Effects of bilingual cooperative integrated reading and composition on students making the transition from Spanish to English reading. *The Elementary School Journal, 99*(2), 153-165.

Eisenstein Ebsworth, M. (2002). Comment. *International Journal of the Sociology of Language, 155/156*, 101-114.

Fleischer-Feldman, C. (1987). Thought from language: The linguistic construction of cognitive representations, in: J. Bruner and H. Haste (Eds.), *'Making sense.' The child's construction of the world.* London: Routledge.

Garcia, E. E. (2002). Bilingualism and schooling in the United States, *International Journal of the Sociology of Language, 155/156*, 1-92.

Goodman, N. (1996). On Starmaking, in: P. J. McCormick (Ed.), *Starmaking, realism, anti- realism, and irrealism.* Cambridge, MA: The MIT Press.

Hamlyn, D. W. (1989). Education and Wittgenstein's philosophy. *Journal of Philosophy of Education, 23*(2), 213-222.

Jepson Green, E. (1997). Guidelines for serving linguistically and culturally diverse young children, *Early Childhood Education Journal, 24*(3), 147-154.

Luntley, M. (2002). Patterns, particularism and seeing the similarity, *Philosophical Papers, 31*(3), 271-291.

Luntley, M. (2003). *Wittgenstein: meaning and judgment.* Oxford: Blackwell Publishing.

Luntley, M. (2004). Growing awareness, *Journal of Philosophy of Education, 38*(1), 1-20.

Marshall, J. D. (1985). Wittgenstein on rules: Implications for authority and discipline in education, *Journal of Philosophy of Education, 19*(1), 3-11.

Peters, R. P. (1966). *Ethics and Education.* London: George Allen & Unwin Ltd.

Porter, R. P. (2001). Accountability is overdue: Testing the academic achievement of limited English proficient students, *Applied Measurement in Education, 13*(4), 403-410.

Reyes, P. & Rorrer, A. (2001). US school reform policy, state accountability systems and the limited English proficient student, *Journal of Education Policy, 16*(2), 163-178.

Rudd, A. (2003). *Expressing the world: Skepticism, Wittgenstein, and Heidegger.* Chicago: Open Court.

Smeyers, P. (1995). Initiation and newness in education and child-rearing, in: P. Smeyers & J. D. Marshall (Eds.), *Philosophy and Education: Accepting Wittgenstein's Challenge* (pp. 105- 125). Dordrecht: Kluwer Academic Press.

Smeyers, P. & Marshall, J. D. (1995). The Wittgensteinian frame of reference and philosophy of education at the end of the twentieth century, in: P. Smeyers & J. D. Marshall (Eds.), *Philosophy and education: Accepting Wittgenstein's challenge*. Dordrecht: Kluwer Academic Publishers.

Spiecker, B. (1977). 'Meedoen en zeker weten' als pedagogische categoric, in: T. Beekman, L. Groenendijk & B. Spiecker (Eds.), *Meedoen en zeker weten: pedagogisch-antropologische opstellen*. Meppel: Boom.

Thompson, L. (2004). Policy for language education in England. Does less mean more? *Regional Language Centre Journal, 35*(1), 83-103.

Vaessen, K., Hoogeveen, K. & Stassen, P. (2003). *De kracht van de kopklas. Basiskwalificatities van de kopklas voor allochtone leerlingen*. Den Haag: Transferpunt Onderwijsachterstanden.

Valdés, G. (2002). Enlarging the pie: Another look at bilingualism and schooling in the US, *International Journal of the Sociology of Language, 155/156*, 187-195.

White, J. (2002). *The child's mind*. London: Routledge Falmer.

Williams, M. (2004). Wittgenstein's refutation of idealism, in: D. McManus (Ed.), *Wittgenstein and skepticism*. London: Routledge.

Wittgenstein, L. (1969a). *The blue and the brown books*. Oxford: Basil Blackwell.

Wittgenstein, L. (1969b). *On certainty* (D. Paul & G. E. M. Anscombe, Trans.; G. E. M. Anscombe & G. H. Von Wright, Eds.). Oxford: Blackwell Publishing.

Wittgenstein, L. (1980). *Remarks on the philosophy of psychology: Vol. 2* (C. G. Luckhardt & M. Aue, Trans.; G. H. Von Wright & 11. Nyman, Eds.). Oxford: Basil Blackwell.

Wittgenstein, L. (2001). *Philosophical investigations* (G. E. M. Anscombe, Trans.). Oxford: Blackwell.

Wright, W. E. (2004). What English-only really means: A study of the implementation of California language policy with Cambodian-American students, *Bilingual Education and Bilingualism, 7*(1), 1-23.

VII
EPISTEMOLOGICAL INSIGHTS AND CONSEQUENCES FOR PHILOSOPHY OF EDUCATION III: DISCURSIVE EPISTEMOLOGY AND THE GROWTH OF KNOWLEDGE

1. Introduction

It seems to be a demand of an acceptable contemporary epistemological approach that it is able to deal with the inevitable uncertainty connected with claims to knowledge. In the foregoing I have argued that fallibilism, as a potential - relatively - 'uncertain' solution that still leaves room for a claim to general validity of knowledge claims, leaves some important questions unanswered, and that a contextualist epistemology might offer a fruitful alternative provided that it is capable of avoiding the pitfall of relativism or arbitrariness. I have demonstrated that a so-called 'discursive epistemology' – in which an interpretation contexts of justification as dynamic-discursive communicative contexts is used – meets these conditions. This discursive epistemological approach suggests that it might not be such a bad thing that we are not able to appeal to something like a relatively solid, deeper-lying 'ground' when it comes to an idea of the justification of knowledge.

Resembling the way in which justification takes place in day-to-day conversations, a discursive epistemology offers an approach to epistemology in which the processes of the development and justification of knowledge is regarded as dependent on an ongoing communicative exchange between speakers and their audience. In such an approach, the legitimacy to make a claim to knowledge – in the definition of being justified to make a claim, or as 'epistemic entitlement' – is not acquired through justificatory grounds that can be regarded as relatively fixed and more 'certain', but rather through grounds that are merely – often implicitly, and mostly for the time being – 'accepted' in the discussion. Understood as such, the safeguard for the epistemological soundness of the claims that are made and the arguments that are used is supplied by the participants in the communication because they are continually watchful as to whether the speaker's contributions meet the prevailing standards within that communicative process. This latter element does, however, raise the question of whether this does not make knowledge too dependent on the more or less coincidental conventions of a communicative community, and whether we then may still be able to speak of the progress of knowledge in a more scientific sense. In this chapter, I will address those questions in connection with the insights gained from the research in chapter six, which dealt with the question of how people

– especially children – learn to participate in the conventions of a communicative practice.

2. The practice-based discursive (re)construction of knowledge

The research in chapter six describes how children's growing into a communicative community also entails their learning to participate in the joint construction and reconstruction of acceptable presuppositions on how the world is; an image that is confirmed by Adler (2008). "Our vast background of beliefs is largely a background of knowledge because of accumulated tacit confirmation that we pick up automatically in acquiring competence in the conversational practice and merely acting on our beliefs", he submits (ibid., p. 348). As such, knowledge is acquired in a holistic way through participation in a communicative process in which the participants, each from his or her own background, continually and actively supply and discuss matters. As participants in the communication, we are socialized as it were in a certain practice-based process of an ongoing reconceptualization of the world. This perspective immediately dismisses the notion of strict, or conservative, conventionalism, because the 'conventions', which seem to be the substratum of our knowledge of the world, are constantly subject to change. By zooming in on the productive, context-transforming role played by the participants in the communication, it becomes clear that the beliefs currently applied within a communicative community do not put any fixed restrictions on whatever is regarded as knowledge within that community. Thus, the danger of conventionalism is averted.

Essential in this, is the idea – which also came to the fore in chapter four – that participation in a communicative process entails not only becoming familiar with a communicative process, but also the understanding that *each* contribution (a pupil's included) implies a proposal to *change* the communicative process. When communicating, speakers are constantly trying their best to put certain topics on the agenda, to introduce new viewpoints or manners of speaking. They try to convince their partners in the conversation to apply certain concepts, appreciate certain matters, use certain methods, etc. As such, each communicative contribution is aimed at changing something – however small – in the way in which the world appears within the communication. This also enables us to explain why practice-based communication may be viewed as an ongoing 'negotiation about how the world is'. That is to say, it is a negotiation or discussion about what we will or will not regard as a 'matter of fact'. In such a process, the currently applied 'conventional' beliefs about the world are, if needed, replaced by 'better' ones – which gives the practice-based communication a criti-cal-discursive nature.

The critical-discursive nature of practice-based communication also shows when we look at the role played by the audience. Besides for the fact that each partner in the conversation, as a speaker, submits proposals to change, as a listener they are in fact always involved in the acceptance, rejection, or discussion of the proposals that are made by other speakers. With every communicative contribution, it will depend on the audience as to whether it is inclined to embrace that contribution to the conversation, or whether they will fight or even ignore it. A speaker who wants to change certain suppositions or standards within the beliefs prevailing in the relevant practice will have to make considerable effort to bring the audience over to their side. However, if the audience sees reason to do so, or is convinced that a communicative contribution will yield a plausible or valuable reconceptualization of the world that fits the other associated presuppositions of the world, it will sooner by inclined to accept that contribution. In the continuing the discussion about how the world is, the contributions made by the partners in the conversations act as the critical touchstone for the way in which the world is described, understood, and discussed. This will not lead to inflexible conventionalism, however, since the way in which the world is understood is constantly tested and reconstructed.

Despite the fact that the presuppositions of the world applied in the communication are principally discussable and replaceable, certain presuppositions or collections of presuppositions may last a long time when they satisfy. Although the communicative context constantly changes, it may seem that no real transformation takes place because certain presuppositions that are deemed to be basic are not put up for discussion and, therefore, continue to function as taken-for-granted presuppositions. At other moments, however, an entire system of beliefs may suddenly topple because a critical contribution to the commu-nication unexpectedly and convincingly undermines a whole collection of presuppositions. This explains, for instance, why, at one moment, knowledge development within certain scientific disciplines seems to quietly drift on within a clearly demarcated paradigm - a situation that Kuhn refers to as a state of 'normal science' – whereas at another moment a true scientific landslide occurs, in which a 'conventional' paradigm crumbles and an alternative paradigm is ready to replace it - which Kuhn calls a state of 'revolutionary science' (cf. Kuhn 1962). This way of interpreting things further makes clear that the relative constancy of some presupposed acceptability standards is not necessarily evidence of a 'more certain' nature of those standards. It only shows that the partners in the conversation have not (yet) seen reason to put the presupposed standards up for discussion, or have not (yet) spotted a possibility for developing a better conceptualization.

When a certain collection of, fundamentally regarded, presuppositions is not put up for discussion for a long time and is consequently applied as self-explanatory, according to Elgin this may be referred to as a "reflective equilibrium" (Elgin 1996, p. 106). Such a state of equilibrium may be regarded as 'reflective', precisely because it concerns an equilibrium that has emerged on the basis of an ongoing negotiation process. "We proceed dialectically", she submits, "[a] process of delicate adjustments occurs, its goal being a system in reflective equilibrium. Achie-ving that goal may involve drawing new evaluative and descriptive distinctions or erasing distinctions already drawn, reordering priorities or imposing new ones, reconceiving the relevant facts and values or recognising new ones as relevant" (ibid., pp. 106-107). After a process of polishing, testing, calibrating, and trying, certain beliefs, standards, methods, etc. will, at a certain moment within the communication, be regarded as "facts". On the same issue, an epistemologist like Brandom concludes that it is in this way that a form of objectivity can also be realized within a dynamic-discursive approach towards justification; an objectivity that does not derive its status from its reference to something like an external reality, or from the application of a specific - rational - method of inquiry, but that is realized intersubjectively within the ongoing communication, and that is normative by nature because it is part of the standards to which the partners in the conversations are bound (cf. Brandom 2000, pp. 196-204, see also: Bransen 2002).

We could now be inclined to explicate the beliefs, standards, methods, etc. that constitute the heart of a 'reflective equilibrium', and take them as a starting point, or norm, for the further communicative process within a certain communicative practice. Apparently, there is a presupposed, a relatively stable, reflectively realized agreement on the validity of certain beliefs and standards, so why would we not explicitly bring that agreement into action as a standard for the further debate? This is, for instance, proposed by Spiecker and Steutel (2001)who advocate a step-by-step method for reaching a 'reflective equilibrium' for the philosophy of education; a method that, in their eyes, could render moral principles that may be deployed for the purpose of evaluating moral judgment. However, the problem of such an explication and anchoring of the prerequisites needed for the reflective equilibrium to arise, or of the elements that are part of the equilibrium, is that it is self-refuting, since the reflective nature of the equilibrium actually demands that disruptions of the equilibrium may occur that could not at all be foreseen and that could also affect the – what we thought – most basic presuppositions (cf. Elgin 1996, p. 133). Therefore, the notion of a 'reflective equilibrium' actually does not tolerate any externally imposed standards of conditions. Each attempt to do so anyway would in fact bring foundationalism back into the picture again, because such standards would be granted an epistemological privilege since they themselves are placed well away from the sphere of influence of the reflection. Such an external intrusion in the equilibrium would also hinder the development of knowledge, rather than encourage it. "Having forsaken foundationalism's reliance on self-justifying claims", Elgin submits, "we can enhance understanding only by drawing on what we have already established" (Elgin 1996, p. 133). In other words, for the further development of our insights we must lean on the insights, standards, and presuppositions as established in the communication so far, without any external foothold. In that sense, the joint development of knowledge has been compared to rebuilding a ship at open sea.

2. Discursive epistemology and progress in (scientific) knowledge

Whatever is accepted as beliefs within a communicative practice, and whatever may, therefore, be considered knowledge within that practice, in a discursive epistemology is made dependent on an ongoing process of negotiation, in which each accepted contribution to the discussion changes, to a greater or lesser extent, the collection of implicitly accepted beliefs and thus the body-of-knowledge used within that practice. That does not, however, clarify what actually might be considered as the 'progress' of knowledge within such an approach. Contrary to change, 'progress' is a normative notion, which means that it may only be referred to on the basis of specific criteria. In scientific discourses, these are usually criteria concerning (research-)methodological justification. A logically consistent and, therefore, conceptually unambiguous argumentation, a theoretical substantiation, or the execution of a controlled experiment may be set as conditions for a valid justification for a claim to knowledge, for instance. From a discursive-epistemological perspective, it can be clarified as to how such criteria for science are, just like all the other acceptability standards together with which they make up the communicative context, are themselves also part of the ongoing negotiation process - and thus also subject to the processes of communicative evolution. As to which criteria for scientific validity are applied may, therefore, also be regarded as the provisional outcome of a critically-discursive change process. The fact that science, then, does not only entail a methodological evaluation of theoretical claims to knowledge, but also the rigorous assessment of the way in which these claims came about, is indeed characteristic of the scientific discourse. For that reason, scientists – according to the Royal Netherlands Academy of Arts and Sciences (*Koninklijke Nederlandse Academie voor Wetenschappen; KNAW*) – are, for instance, not only supposed to conduct research in line with the currently accepted scientific methods, but must also describe their methods "in such a way that others are able to assess their validity" (KNAW et al. 2001, p. 5). It is most probably thanks to the trust in the critical, self-corrective abilities of science that we also assign science the role of knowledge authority within our society.

In order to gain insight into what all this means for individual scientists, we can now use the insights that chapter six rendered in respect of understanding the notion of 'expertise'. Concerning partici-pation in a communicative practice – such as a scientific communicative practice – it is submitted in this chapter that 'expertise' has a twofold nature. The term 'expertise' refers to the extent to which a participant in the communication is familiar with the collection of presuppositions that are shared within the communication. Obviously, this is also important when it comes to scientists. Scientific expertise presupposes matters like knowledge of a certain academic discipline, being able to design a consistent line of argumentation, the competence to apply certain methodological principles, etc. At the same time, 'expertise' also invol-ves a speakers' ability to – successfully – put parts of the communicative context up for discussion. Here, the 'expertise' might well lie in the ability to question those suppositions that others take to be basic (after all, the questioning of a random part of the context is reserved for each participant in the communication, including the 'novice'). Given all this, the expert scientist would appear to not only be someone who excels in the conducting of research in line with the general idea of 'proper science', but also someone who, at the same time, is able to critically, and successfully, question certain principles that are deemed to be at the core of the scientific endeavor. Excellent scientists, according to Elgin, are scientists who at the least have the ability "to operate successfully within the constraints the discipline dictates", but are also able "to challenge those constraints effectively" (1996, p. 123).

3. Consequences of a discursive-epistemological approach for philosophy of education

The foregoing shows that epistemic criteria do not precede scientific discourse, but are part of the ongoing scientific discussion. This supports the thought already proposed in chapter two that the idea of the 'the primacy of epistemology' must be abandoned. This insight bears also on the philosophy of education. We do not need to be concerned, however, that this would imply the inevitable downfall of strict epistemic criteria for doing philosophical-educational research. It only shows that the criteria for what is considered to be a philosophically acceptable contribution to the debates in philosophy of education are themselves a topic of discussion. This does not only involve the question of what is considered an acceptable method of inquiry. The tasks of the philosophy of education, for instance, are also part of the discussion - as is illustrated in this dissertation. The things that the philosophy of education should concern itself with, what it can do, and how it should go about, is simply part of the ongoing negotiation on what the world - in this case, the world of philosophy of education - is like. With each contribution, participants in educational-philosophical debates inevitably bring up for discussion the beliefs, criteria, or starting points that were previously taken as taken for granted. At the same time, they act as guardians of the limits for what may apply as philosophically acceptable contributions to that debate. In this connection, my reaction to fallibilism in the philosophy of education may be seen as illustrative. On the one hand, I question a certain standard that seems to be broadly taken for granted by philosophers of education in the field. For I suggest to abandon the idea that for the justification of claims within the philosophy of education, beliefs are needed to which a certain epistemic certainty is attributed. At the same time, the reaction acts as a defense of another standard to which, as we have seen, usually an important role is assigned in academic research – hence, also in the philosophy of education (cf. chapter one). For I argue for the rejection of attributing epistemic certainty by, among other things, pointing out that it erodes the reflectivity of the philosophy of education, thereby endangering its critical, self-correcting capability as an academic discipline.

How the philosophy of education should be designed is, therefore, now made dependent on the educational-philosophical discourse itself. In separate contributions, proposals are made to accept or reject certain acceptability standards for the philosophy of education. In light of other acceptability standards, those proposals are then incorporated, or not. The content of philosophy of education thereby seems concurrently dependent on what the separate speakers are committed to, and on the commitments that play a role in the wider educational-philosophical discourse. In any case, commitment seems to play an important role when it comes to the format and content of the philosophy of education. That is something we have seen before in this dissertation. In chapter two I showed that, in reaction to the rejection of the 'primacy of epistemology', contemporary antifoundationalist philosophers of education suggest to take a basic commitment – in their case, to fighting exclusion – as a starting point for philosophy of education. Although I have argued that, in light of my findings, such a proposal does not seem very fruitful, there is now again reason to believe that there might well be a shift taking place within the philosophy of education from a 'primacy of epistemology' to a 'primacy of engagement'. However, whatever such a 'primacy of engagement' might entail exactly is not at all clear as yet. This is reason to further investigate the idea of the 'primacy of engagement' in the next chapter.

4. References

Adler, J. E. (2003). Knowledge, truth, and learning, in R. Curren (Ed.), *A companion to the philosophy of education* (pp. 206-217). Malden: Wiley-Blackwell.

Adler, J. E. (2008). Conversation is the folks' epistemology, *The philosophical forum, 39*, 337-348.

Brandom, R. B. (2000). *Articulating reasons: an introduction to inferentialism.* Cambridge, MA: Harvard University Press.

Bransen, J. (2002). Normativity as the key to objectivity: An exploration of Robert Brandom's articulating reasons. *Inquiry, 45*, 373-392.

Elgin, C. Z. (1996). *Considered judgment.* Princeton: Princeton University Press.

Koninklijke Nederlandse Akademie van Wetenschappen, Vereniging van Universiteiten & Nederlandse Organisatie voor Wetenschappelijk Onderzoek (2001). *Notitie wetenschappelijke integriteit. Over normen van wetenschappelijk onderzoek en een Landelijk Orgaan voor Wetenschappelijke Integriteit (LOWI).* Amsterdam: Auteurs.

Kuhn, Thomas (1962). *The Structure of Scientific Revolutions.* Chicago: The University of Chicago Press.

Spiecker, B. & Steutel, J. (2001). Reflective equilibrium as a method of philosophy of education: justifying an ethical conception of children's sexual rights, in: Frieda Heyting, Dieter Lenzen and John White (eds.), *Methods in philosophy of education* (pp. 30-43). London: Routledge.

VIII
THE PRIMACY OF COMMITMENT IN PHILOSOPHY OF EDUCATION

1. Introduction

Developments in epistemology undermine the traditional ambition of philosophy of education to contribute to the formulation and justification of basic principles for educational practice (Snik et al, 1994). Particularly antifoundationalism, which rapidly wins ground, makes this ambition seem too pretentious. In philosophy of education, antifoundationalism appears to cause a tendency to restrict the practical ambitions of this discipline to demonstrating the (individual or social-contextual) restrictions of any attempt to come to 'justified' foundations (Van Goor et al., 2004). By drawing attention to the limited validity of any justification, these authors want to put the powers of philosophical judgments into perspective, hoping to resist exclusion of alternative positions and to keep an open mind to their possible relevance. In order to achieve this, they stress the importance of formulating 'counterpractices' (Biesta, 1998; Peters & Lankshear, 1996), of 'deconstructing' existing positions (Biesta, 2001; Egea-Kuehne, 1995), or of analytical applications of irony in philosophy (Van Goor & Heyting, 2006). These approaches characteristically bring established answers up for discussion, rather than providing new answers.

Though being aware of potential exclusion and the possibilities of resisting it unmistakably shows evidence of a normative orientation, the above-mentioned positions cannot be called normative in the traditional, prescriptive, sense, because the desirability of avoiding exclusion is not primarily brought up and defended as a position in its own right. The authors concerned especially stress the impossibility of determining even relatively certain foundations, which makes that the philosopher of education is ultimately thrown upon his own preferences, opinions, and affinities - with all their limitations. In short, with respect to knowledge and knowledge-claims these authors seem to replace the traditional 'primacy of epistemology' with the 'primacy of commitment' as I will call it. This way of looking at knowledge-claims implies that positions are not considered reducible to such foundations or basic principles that can show epistemological privilege; instead they are seen as ultimately dependent on the author's specific engagement with the situation that I will refer to as 'commitment'. The specific interpretation of this commitment and its consequences for philosophy of education will be the subject of my analysis.

Though most antifoundationalist authors seem to agree that avoiding exclusion does have practical relevance, it does not solve all problems with respect to philosophy of education's practical relevance. For example, if philosophy of education should restrict its pretensions to making critical analyses in order reveal excluding one-sidedness, which could break discussions open, how should we evaluate such critical analyses in their turn? Doesn't this result in a paradoxical situation, because these analyses themselves would be liable to similar restrictions, and result from the commitment of its author (Roth, 1995)? What kind of relevance could such a philosophy of education still have? Doubts like these make some authors concerned for an impending "intellectual paralysis" (Blake et al, 1998, p. 5), which according to Smeyers (2005) should be counter-acted by reintroducing some kind of explicit normativity in philosophy of education. If educational thinking inevitably gives evidence of the commitment of its originator, he seems to argue, we better make this commitment explicit from the beginning. That would make the position of the author clear for everyone.

Apparently, it's not at all clear what the 'primacy of commitment' exactly implies, and how we should understand its consequences for philosophy of education. In order to answer these questions, I will first consider the interpretations of 'commitment', or 'being committed' (2), and the related interpretations of the 'primacy of commitment' (3). A further inspection of the latter brings to light a discrepancy with the non-relativist ambitions of the antifoundationalist authors (4), which gives rise to a reconsideration of the interpretation of commitment and its role in the development and justification of knowledge (5). Against this background, I develop a more precise interpretation of the 'primacy of commitment' that enables us to deal with the reproach of relativism (6). In conclusion, I discuss some consequences of my findings for under-standing the practical relevance of philosophy of education (7).

2. The concept of commitment

The concept of 'commitment' has been differently defined by different authors, sometimes including the behavior commitment finds expression in, and sometimes only consisting of the attitude that lies at the basis of what is considered committed behavior. Meyer en Allen (Meyer & Allen, 1991, p. 62), for example, distinguish 'attitudinal commitment' from 'behavioral commitment' whereas Morgan (2005) mentions both as two characteristics of commitment. However, there seems to be agreement about two aspects that are considered to be of crucial importance to the concept of commitment. As Becker (1960, p. 35) in his much cited analysis concludes, 'commitment' refers to an affectional situation of 'being committed' - which I will call the 'feeling-dimension' of commit-ment - on the one hand, and the activity of making a commitment to a specific content - which I will call the 'content-dimension' of commitment - on the other, leaving in the middle whether it should be accompanied with some kind of physical behaviour. Willms, for example, defines stu-dents' commitment to their schools - apart from the behavioural compo-nent - as "pertaining to student's sense of belonging at school *and* accep-tance of school values [italics added]" (2003, p. 8). In this example, the sense of belonging represents the feeling-dimension, and the acceptance of school values represents the content-dimension.

The necessary presence of both of these two dimensions, concer-ning feeling and concerning content, recur time and again (Bellah, 1985; Blustein, 1991; Echeverria, 1981; Lieberman. 1998; Lovie, 1992; Montefiore, 1975; Morgan, 2001: Sternberg. 1987). For example, Blustein stresses that the presence of the feeling-dimension is not enough to speak of commitment. because "not everything or everyone people care about is an object of commitment"; there has to be some kind of evaluative judg-ment as well (Blustein,, 1991, p. 120). Lieberman stresses the other side of the picture. According to him one cannot speak of commitment if there is only an endorsed content, and not the necessary affectional attachment to it. As a consequence, serious doubts concerning the endorsed content can-not occur in typical cases of commitment (Lieberman, 1998, pp. 40ff.).

Commitment, then, seems a matter of feeling involved in certain - often political or cultural - problems or situations, accompanied by taking a specific position as regards content, i.e. the cause one is committed to. Consequently, I will broadly define 'commitment' as 'an appraising invol-vement in a situation'. The 'appraising-part refers to the content-dimen-sion, and the involvement'-part refers to the feeling-dimension; in 'com-mitment', both dimensions are present. The kind of situation a person is involved in does not make any difference for the necessary presence of both dimensions. Both are there, whether one is thinking of commitment to institutions (Coser, 1974), religions (Audi, 2000), communities or practices (Bellah, 1985), organizations (Becker, 1960; Buchanan, 1974), activities (Lieberman, 1998; Scanlan, 1993), and relationships (Sternberg, 1987). For example, like Willms (2003) with respect to school commit-ment, Buchanan (1974) considers 'organizational commitment' characteri-zed by affectional attachment to an organization as well as endorsement of its aims and values. Bellah et al. discuss commitment to communities, which requires feeling "patterns of loyalty and obligation" as well as en-dorsement of what "its hopes and its fears are, and how its ideals are exemplified in outstanding men and women" (1985, p. 154).

A more profound understanding of the concept of commitment

will require a more detailed view of the interpretation and mutual relations of both characteristic dimensions. Current philosophical interpretations of commitment especially appear to diverge with respect to the way both dimensions are related, one group of authors attributing priority to the feeling- dimension over the content-dimension and a second group of authors attributing priority in reverse. In the former case, the feeling-dimension is considered a derivative of the content the person concerned endorses; in the latter case the content-dimension is considered to ensue from the affectional involvement, i.e. the feeling-dimension. The differ-rence between both interpretations can be exemplified with two phases in the history of Sartre's thinking.

In his existentialist Opus Magnum '*Being and nothingness*', Sartre (1969) relates commitment to the human ability of taking life in one's own hands. The present moment of choosing is of vital importance to Sartre's existentialist conception of commitment (Craib, 1976, p. 29). In his existentialist phase Sartre primarily relates commitment to the existential moment in which a human subject makes his own choices in specific situations, manifesting his personal authenticity. This authentic position-finding of the subject takes place in the direct personal and situational existence, and precedes the restrictions that are inherent in any social convention. In this way, Sartre's existentialist interpretation of commit-ment gives priority to the feeling-dimension. The content of the position the individual chooses ensues from an existential moment in which the individual is purely left to his own devices, and not from a consideration and application of some endorsed value system. As a consequence, the position a person chooses at any particular moment is no predictor of what positions the subject will choose in future situations. Any present choice is, as David Cooper formulates it, "a 'fundamental choice' for which no justification is available" (1990, p. 143).

Whereas in Sartre's existentialist phase the appraising orientation of the subject resulted from making a fundamentally new choice at every occasion, in his later Marxist phase Sartre displayed a quite different view of commitment. Now he proves himself a fanatical follower of a political doctrine that considers commitment following from adherence to a theory, a content that fundamentally restricts future choices. De Brabander (2003) explains this radical turn in Sartre's views as a sign that Sartre now identi-fies with the picture of the committed author who abandoned the dream of being an impartial and 'free' choosing member of society as being unre-alistic (cf. Sartre, 1948, pp. 76ff.). His 'conversion' to Marxism made Sartre an exponent of a conception of commitment that prioritizes content, and that explains the affectional attachment to a choice from the fact that it is a consequence of the specific value system the person endorses.

Those philosophers of education, who replace the primacy of epistemology with the primacy of commitment (cf. chapter two), also give evidence of a conception of commitment that either prioritizes the content-dimension or the feeling-dimension. The former group will consider the appraising involvement in a situation primarily a consequence of the belief-system (ideas, values, etc.) that determines what positions in what situations an individual can get enthusiastic about. However, these authors do not associate this belief-system - this prior content - with an explicit and crystallised doctrine like Marxism. These philosophers of education rather consider this prior content as extracted from the social-historical context of the individual - a context that is either interpreted in terms of group-membership Fitzsimons & Smith, 2000; Giroux, 1997; McLaren & Giroux, 1997; Peters & Lankshear, 1996; Weinstein, 1995), or in terms of discursive practices like discourses, language games, or social-communicative systems Biesta, 1998; Blake et al, 1998; Heyting, 2001; Marshall, 1996; Peters, 1995a; Peters & Marshall, 1999; Ruhloff, 2001; Simpson, 2000; Usher eta!, 1997). In the first case, the content that constitutes the primary source of commitment consists in the views, beliefs, ideals and values as shared by the members the group, and consequently the commitment of an individual will reflect the norms of the group to which he belongs. In the second case, the content that constitutes the primary source of commitment consists in the system of beliefs that is taken for granted in the discursive practice involved. Citing Foucault we can say that "each discursive practice implies a play of prescriptions that designate its exclusions and choices" (1977, p. 199), which produces the content that is considered the primary source of commitment.

There is also a group of antifoundationalist philosophers of education who consider the appraising involvement in a situation as ensuing from a primarily affectional process. These authors prioritise the feeling-dimension of commitment. In their view, something like 'being touched' in de specific situation is constitutive of the way the content-dimension of commitment takes shape, instead of the other way mound. This affectional source of commitment is sometimes conceived of as strictly individual and personal in nature, whereas others rather consider it to be relational in nature. Kohli (1998) and Prior McCarthy (1995), who see the biographical constitution of the body as the final source of position-taking in the world, take the individual-oriented stance. In their view, the physically experienced accumulation of life-events causes a specific, pre-reflexive way of being, from which the contents of individual appraisals result.

Noddings (1984) represents a relational view of the affectional process that is assumed to be at the root of commitment where she stresses the 'natural' and spontaneous response of care that any confrontation with another person can evoke. Lynda Stone (1995) also implies an inter-personal affect as the source of commitment, but she explains it in terms of friendship. Still others see the interpersonal affect that constitutes the source of commitment simply in being touched by the 'otherness' of the other person (Biesta, 1999; Child et al, 1995; Masschelein, 1998; and 2000; Säfström, 1999). Whatever form it thought to take, these authors consider the interpersonal affect to be the source of commitment, of the appraising interpretation of the situation by the subject. Noddings, for example, formulates it as follows: "The 'ought' - better the 'I ought' - arises directly in lived experience. 'Oughtness', one might say, is part of our 'isness'. [...} I have called this spontaneous response 'natural caring' (1998, p. 187). In other places, she reacts against Kantian views, because they would priorities content (principles) over feelings: "An ethic of care inverts these priorities. The preferred state is natural caring; ethical caring is invoked to restore it. This inversion of priority is one great difference between Kantian ethics and the ethic of care" (Noddings, 1995, p. 138).

Summarizing I can say that authors give priority to either the content-dimension or the feeling-dimension of commitment. Authors who give priority to content usually consider this content as resulting from a context, - either a group, or a discursive practice. Authors who give priority to feeling consider this affectional source of commitment to be either an expression of physical individuality, or of interpersonal relatedness.

3. The primacy of commitment (I)
The various interpretations of the concept of commitment also affect the way authors understand the antifoundationalist primacy of commitment with respect to knowledge claims, and the practical consequences they ascribe to it. For example, authors who consider the group-context the primary source of commitment, will ultimately consider knowledge-claims as reducible to shared beliefs, ideals, preferences, and values of the group. Accordingly, these authors will also be inclined to stress the social risks that this state of affairs implies. If knowledge claims reflect belief-systems of the group, they argue, such claims will also reflect current power relations and group interests, which would play into the hands of dominant group members and exclude members of other groups. Consequently, authors who understand the primacy of commitment as related to group-contexts will be inclined to associate the development and justification of knowledge with social struggle. For example, Henry Giroux argues that science and rational procedures in general should be understood as part of a broader historical, political, and social struggle" (1997. p. 195).

If a discursive practice is considered to be the context. that determines the content- dimension of commitment, justification of knowledge is ultimately seen as reducible to the belief-system that is implicitly taken for granted in a specific discourse or language game. Such an interpretation of the primacy of commitment makes justification only convincing within the boundaries of this discursive practice. For example, Michael Peters argues that the validity of any claim is subject to the restrictions set by the rules of the language game in the context of which the claim was made, because "...the rules are irreducible and there exists an incommensurability among different games" (Peters, 1995, p. 391). To claim any authority outside of the boundaries of the current language game would require playing another language game according to Peters. This interpretation of the primacy of commitment draws attention to the social phenomenon of communicative boundaries and exclusion. In both context-related interpretations of the primacy of commitment, commitment is determined by a contextual belief system that functions as a boundary beyond which no justification is possible, and in both cases authors stress the importance of being aware of the restrictive effects of this situation in the social realm. Consequently. these authors stress the importance of trying to overcome the conventional belief systems that cause these restrictive effects.

Those authors who give priority to the feeling-dimension of commitment interpret the primacy of commitment depending on the kind of feeling they consider fundamental, which also explains the kind of practical consequences they primarily stress. However, in all cases, justification is ultimately considered dependent on the affective condition of the person(s) concerned, and not on a conventional social or discursive context. As Noddings sees it, the reasons we can and do give for our actions often "point to feelings, needs, impressions, and a sense of personal ideal rather than to universal principles and their application" (1984, p. 3). Prior McCarthy, who accentuates the individual nature of feeling as the source of commitment, writes: "In speaking and knowing, it is not rules that are crucial, but the knowers and speakers themselves" (1995, p. 45). Though recognising expertise as being able to apply certain rules, she alerts to the limited scope of all rules, which puts individual judgment in the forefront: "But, sooner or later, the questions are left to the judgment of individuals" (ibid., p. 37). According to Prior McCarthy, expertise ultimately depends on the question whether "he or she possesses a certain kind of past life, a particular kind of biography. What I mean by 'biography' is a life itself and not a narrative or a set of ideas" (ibid., p. 36).

Though considering the feeling-dimension the primary source of commitment, these authors are well aware that processes of giving and accepting reason are regulated by social and communicative conventions. Like authors who consider contextually determined content the source of commitment, these feeling-oriented authors stress the restrictions of our conventional instruments for formulating and testing arguments. Exactly for this reason they consider it important to alert to the affectional source of commitment. In their view, taking this affectional source seriously is the obvious way to escape from the restrictions of these social conventions and get our thinking better attuned to our basic feelings. To many authors, this possibility of escaping from conventions gets an ethical import as well. Noddings (1995) is very explicit in this matter. According to her, the spontaneous, affective response of 'natural care' should be the starting point as well as the aim of ethics. In her view, no genuine relation of care can result from any ethical doctrine; if anything, ethical doctrines should serve the blooming of the original caring-affect. Kohli argues in a similar way. According to her, the original feeling resides in the bodily situatedness of the subject, and the almost absolute supremacy that was traditionally ascribed to reason - and to the related importance of generalization - has resulted in the neglect of views that stem from the physical-personal dimension of existence. In order to emancipate from this tyranny of reason, Kohli wants "to put voice to the production of my own body" (1998, p. 519).

Similar arguments can be found in authors who locate the basic feeling in being confronted with the 'otherness' of other people. They also consider this interpersonal affectional source of ethical importance because it can free from conventional 'blinkers'. This original interpersonal affect is considered the source of accepting responsibility for the other as other, a responsibility that precedes principles and beliefs, and that is a natural part of being in a situation. According to these authors, it is impossible for a subject to withdraw from this pre-reflectively anchored responsibility, because the subject is only constituted as such in this relation (Biesta, 1999; Säfström, 1999). It is important to recognize, that claiming and justifying knowledge can ultimately be reduced to this affective-ethical situatedness of the subject. "For, in this light", Standish writes, "the ethical is there at the start in our actions and projects, and hence inevitable there at the start in research in education" (2001, p. 498).

All in all, the various conceptions of commitment result in two interpretations of the primacy of commitment with respect to knowledge. The primacy of commitment means that claiming and justifying knowledge is considered either to depend on the belief system that characterises the current social-communicative context, or on the pre-reflective, affectional being in the situation. Both interpretations of the primacy of commitment share some crucial elements. Across the board authors consider the conventional nature of claiming and justifying knowledge a limitation that should be overcome in order to combat exclusion - either of alternative social-communicative contexts, or of original, pre-reflective ways of feeling as sources of judgment. In all cases, the unquestioning acceptance of prevailing conventions is to blame for excluding effects. That raises the question how these authors imagine such prevailing conventions to function, and in particular how they conceive of the possibility of escaping from such conventions.

4. How to escape from conventional restrictions

The restrictions that result from the social-conventional embedding of knowledge and its justification appear to pose a problem antifounda-tionalist philosophers of education want to overcome. At first sight, this idea of the social embedding of knowledge may cause a semblance of relativism to these interpretations of the primacy of commitment (Williams, 2001, p. 221). However, the authors in my sample deviate from relativists in one important respect. Relativism not only implies making the acceptability of claims dependent on the locally current set of norms, it also implies the impossibility of bringing these norms up for discussion (ibid., p. 226). Antifoundationalist philosophers of education appear not to share the latter. They rather explicitly aim at overcoming or at least challenging such local restrictions. Consequently, they do not consider it impossible to escape from the compelling character of criteria for acceptability that are part of conventional communicative contexts (Van Goor et al., 2004, p. 185). In this respect, they deviate from relativism. This raises the question: how do they imagine the possibility of such an escape?

A first view of how to escape can be reconstructed from texts that prioritize contextually determined content as a source of commitment. Peters (1995b), who considers language games the determining contexts, thinks we can escape from its restrictions by making use of the differences between language games that exist next to each other, - each of them characterized by its own specific norms for justification. This "multiplicity and proliferation of forms of reason, defined by the rules of particular discourses or language-games" (ibid., p. 391) makes it possible to escape by changing language games. Changing language games also would make it possible to discuss one language game from the perspective of another language game.

This impression of things can be found in many of the discussed antifoundationalist philosophers of education. Often arguing from a critical-pedagogical perspective, they stress the possibility of questioning the socially prevailing knowledge frame by confronting it with alternative frames - practices (Biesta, 1998), or narratives (Giroux, 1997; Peters & Lankshear, 1996). For example, Gur-Ze'ev mentions the importance of 'counter-education' that could offer "possibilities for identifying, criticizing, and resisting violent practices of normalization, control, and reproduction" (1998, p. 463). Such alternative frames, discourses, or language games that seem a necessary precondition for the possibility of escaping from prevailing ones, do not necessarily have to exist already. Some authors also mention the possibility of creating them (Biesta, 1998; Giroux, 1997; Gur-Ze'ev, 1998; Peters & Lankshear, 1996). Especially this last suggestion seems surprising because these same authors also stress the restrictive nature of current contexts. How is it possible to create a new discourse, language game, or belief system if one is confined to the restrictions of an existing one?

For similar reasons, it is hard to imagine how it would be possible to change between existing vocabularies. Being confined within the limits of a specific communicative context seems to imply the impossibility of imagining alternatives, in particular if they are conceived as mutually incompatible as Peters (1995b, p. 388; and p. 91) explicitly does and other authors at least implicitly suggest. Either a communicative context keeps one caught within its limits - thus making acceptability and justification restricted to the current context -, or the conventional context is open and allows for change and revision of positions - but in that case the context would no longer be restrictive. To have both at the same time seems impossible. Antifoundationalist philosophers of education do not extensively discuss how the restrictive nature of communicative conventions can be consonant with criticizing and changing them, but some ideas can be found in Rorty.

Rorty, who uses the term 'vocabularies' for communicative contexts, describes the role of communicative conventions just like the above-mentioned philosophers of education. Like them, Rorty considers any use of language - including the norms that regulate processes of claiming and justifying knowledge - to be an expression of a specific vocabulary that is characterized by a specific set of concepts, beliefs, and norms (1989, pp. 11 ff.). In addition, Rorty does not think either that we are irrecoverably at the mercy of a current vocabulary, even though we may be more familiar with some vocabularies because of socialization. In his view, the influence of socialization explains why people may be inclined to unquestioningly hold on to a current vocabulary. Just like the antifoundationalist philosophers of education, Rorty thinks it possible to change vocabularies and to choose alternative ones. Even our 'final vocabulary' can be reconsidered, resulting in a redefined self. "We redescribe ourselves, our situation, our past in those terms and compare the results with alternative redescriptions", Rorty writes (ibid., p. 78). Finally, Rorty also thinks it possible to create new vocabularies, although he doesn't think that it is easy.

Discussing how it is possible to create new vocabulary. Rorty also seems struck by the riddle how a vocabulary can be restrictive and closed, and open for change or revision at the same time. He stresses the exceptional nature of this phenomenon, and - speaking of the inspiration and genius of a poet - he also stresses the special abilities that are required to make this happen (ibid., p. 12). However, Rorty's explanations of this phenomenon are not so clear. His reference to the inspiration of a poet suggests the possibility of relating to a vocabulary, which seems at odds with Rorty's view that we always find ourselves in a vocabulary, subjected to its restrictive effects (ibid., p. 75). At this point, the main problem we seem to be facing, is how to find a position outside any vocabulary. Rorty does not solve this problem, and the context-oriented philosophers of education leave it in the dark as well. But Rorty's use of the words 'inspiration' and 'genius' suggest that a solution might still be found by considering the personal, affectional sphere that is not pre-formed by vocabularies, as those philosophers of education, who prioritize the feeling-dimension of commitment also suggest. In their view, this affectional dimension makes it possible to escape from being confined to communicative conventions.

Masschelein most explicitly discusses this 'critical' function of the feeling-dimension. He considers it a human capacity "that is 'older' than the order of representation" (Masschelein, 1998, p. 528). Because this affect precedes any form of linguistic expression, Masschelein considers it the ideal position for recognizing the defectiveness of any set of linguistic instruments. In his view, we have to be sensitive for what precedes linguistic conventions, because "at such a moment the linguistic attempt to understand and to grasp reality is seen for what it is, precisely because it fails" (ibid.). At the level of feeling, it becomes possible to get to the very heart of things, and to gain an idea of the shortages of our linguistic representations. Other authors ascribe a similar function to the original affect, like the response of caring in Noddings, or bodily being in Kohli. From such affectional positions, the restrictions of conventional linguistic instruments are thought to become visible. However, I doubt whether this movement 'back to the affectional roots' suffices to explain the possibility of bringing the belief systems of conventional contexts up for discussion. Even if we consider such affectional states as pre- or extra-linguistic, the problem remains how one would be able to express an insight resulting from such a state in a way that would be understandable for other people. Trying to do so would throw us back to the confinement of a conventional vocabulary. Naming such affectional roots - e.g. as 'care' (Noddings) or as 'friendship' (Stone) - would already be a problem. The language of emotions can profoundly differ from culture to culture. Nature, classification, meaning, and appreciation of emotions, as well as ideas about what situations call for what emotions, about the power that is ascribed to emotions and the ways to deal with them can diverge extremely (Heelas, 1996). Antifoundationalism, which forbids any 'direct' entry to external reality - as a result of which no epistemologically privileged knowledge seems possible - also forbids the possibility of 'directly' understanding and representing feelings.

Rorty recognizes this problem, and he rejects any appeal to 'original' affects as standards for evaluating communicative conventions. As he formulates it: "... the claim that an 'adequate philosophical doctrine' must make room for our intuitions is a reactionary slogan, one which begs the question at hand. For it is essential to my view that we have no prelinguistic consciousness to which language needs to be adequate, no deep sense of how things are which it is the duty of philosophers to spell out in language" (Rorty, 1989, p. 21). Consequently, the interpretation of the role of genius in criticizing conventional vocabularies as an appeal to any kind of pre-linguistic origin must be rejected as well.

It appears to be rather complicated to realize the ambition of maintaining the primacy of commitment with respect to knowledge while rejecting epistemological relativism. Trying to do so confronts us with two problems, depending on the interpretation of the concept of commitment. Either the idea that the communicative contexts people participate in determine the preferences, opinions, and affinities that lie at the basis of knowledge-claims, makes it difficult to understand how it is possible to criticize this same context. Or the idea that an 'external' position, apart from communicative conventions, would be possible and enable us to judge conventions, seems to contradict the inability to express such judgment in any understandable - and therefore conventional - terms. It seems that one is either confined to current conventional belief systems, or to one's own pre-linguistic affect. A non-relativist interpretation of the primacy of commitment will require reconsideration of some of the concepts it involves in such a way that getting epistemological privilege back in by the backdoor by creating any kind of direct access to reality is avoided, while at the same time allowing for bringing contextual belief systems up for discussion. As the main problem seems to be caused by the role that is ascribed to communicative contexts as related to commitment, a reconsideration of both concepts and their mutual relatedness might offer a solution.

5. Restrictions versus conditions of possibility

The function of commitment in developing and justifying knowledge as suggested in the discussed texts, seems primarily to indicate a situation of confinement, either to the belief systems and criteria of communicative contexts, or to the 'insights' of a pre-linguistic affect.

This restrictedness does exclude any epistemological privilege - as antifoundationalists maintain -, but it also seems to exclude possibilities of overcoming these restrictions - as these antifoundationalists' non-relativist ambitions would require. This situation seems to result in a stalemate. However, a closer look at the way authors practically deal with this problem in their argumentations reveals a quite different picture. These antifoundationalist authors devote a substantial part of their argumentations to urge people with deviant belief systems or ways of feeling on to surmount restrictions, and they do so taking refuge in arguments. In other words, they operate as if arguments could be understood and accepted across the boundaries of communicative contexts, or as if arguments could change original affects that make people choose specific positions with respect to content.

Noddings, for example, who prioritizes the feeling-dimension of commitment, explicitly recognizes the consequence that ultimately convincing others cannot be a matter of giving reasons, but has to take place in the affectional realm instead. Against this background, teaching care will not consist in passing on rules, but in "nurturing an ideal" (Noddings, 1984, p. 124). According to this way of seeing things, people can only change position if a change of affect has taken place first. Despite of that, Noddings does not behave like that when she addresses her main opponents, the adherents of a principle-oriented Kantian ethics. She makes use of arguments as her main weapons instead, and she even seems to be aware of the anomaly of this approach as compared to her own theory: "I shall have to argue for the positions I set out expressively" (ibid., p. 2).

At first sight, Kohli seems to proceed in accordance with her affect-oriented theory. She localizes the affectional basis of commitment in the bodily-biographical situatedness of the subject. Reacting to Peter McLaren and Ilan Gur-Ze'ev, she characterizes McLaren's argumentation as rooted in his "immediate connection to the oppression of Chicano/Chicanas in California and Mexico". Gur-Ze'ev's as an expression of "the politico-philosophical restraint that comes from an Israeli whose memory of the holocaust ... is still painfully fresh" (Kohli, 1998, p. 511). However, she doesn't leave it at that. She continues to bring up arguments in order to convince these authors of her views.

This way of proceeding makes it hard to maintain, that the

affective dimension should be considered the ultimate source of our positions and beliefs. Consequently, the concept of commitment cannot be understood exclusively prioritizing the feeling-dimension as the source for renewed content. Accordingly, this feeling-dimension cannot be understood as a pure pre- or extra-linguistic position either. If affects can be influenced by means of linguistic arguments - a presupposition these authors demonstrate in their texts -, or indeed if affect can cause new content, there has to be some kind of exchange between affect and content. Even Masschelein, who, unlike Noddings, Kohli, and Stone. does not specify a particular type of affect as the ultimate source of content, still urges to take refuge to affective experience for genuinely renewing content while using arguments as his main instrument.

This does not mean, that one should prefer the alternative interpretation of commitment, which prioritizes content over feeling, because this interpretation still leaves in the dark how the barriers of communicative contexts can be overcome - considering that the validity of arguments is restricted to the current context. Consequently, the content-oriented interpretation of commitment does not allow for a non-relativist interpretation of the primacy of commitment either. After all, if commitment - now resulting from endorsed contents that determine for what causes someone can become enthusiastic - is determined by the communicative context, what could cause the readiness to start playing a different language game, as Peters (1995b) calls it? To explain that, again, we have to accept that there is some kind of two-way traffic between the feeling-dimension and the content-dimension of commitment. This results in the first conclusion: a non-relativist interpretation of the primacy of commitment requires a concept of commitment in which both dimensions - the feeling of 'being committed' and the content one makes a commitment to - go hand in hand, are mutually related, not prioritizing either dimension.

A second conclusion that can be drawn is that the function of - both dimensions of - commitment with respect to the development and justification of knowledge cannot be understood as an insurmountable barrier for accepting alternative views. The appraising involvement in situations is changeable, or is at least treated as such. This changeability even seems the main objective of most authors, who primarily seem to try and convince others of the acceptability of their own position. As demonstrated above, authors who prioritize the feeling-dimension of commitment demonstrate this impulse to convince others with the help of arguments authors who prioritize context-related content act likewise. For example Peters, who wants his modernist readers to replace their language game with a postmodernist one, sets out to bring this about by means of arguments. "First and foremost, such a philosophy would involve a serious engagement and re-evaluation of modernity ... It concerns itself with deconstructing and providing a genealogical critique of the foundational interpretative frameworks which have served to legitimate techno-scientific and political projects in the modern world" (Peters, 1995a, p. 203). Though a protagonist can be expected to demonstrate some resistance against abandoning his current position, treating his adherence to this position as absolute will drive us into a pragmatic contradiction.

A non-relativist interpretation of the primacy of commitment will require a conception of commitment in which the feeling-dimension and the content-dimension go hand in hand, and which does stress the willingness to persuade others across barriers instead of the inability to recognize and accept alternatives beyond the current context or affectional basis. As a result, commitment can be defined as an appraising involve-ment in a situation that is characterized by a mutual dependence of the affect- and the content-dimensions , and that motivates people to propagate their position amongst others. This view of commitment is, for example, expressed by De Vries, who describes a committed intellectual as someone who "... ascends the stage, gets up to speak, places a signature on a reader's letter ..., publicises a burning j'accuse, joins a talkshow. What has to be defended is the truth, or civilisation, or whatever should pass for that ... [Trans.: RvG] (De Vries, 1992, p. 111).

This enthusiasm seems hardly imaginable without assuming that argumentations can in principle be accepted outside the current context, an assumption that is required for a non- relativist interpretation of the primacy of commitment. This means, that such a non-relativist inter-pretation requires a revision of the concept of context as well. Under-standing commitment as the willingness to propagate the own views in the presence of a dissenting forum will also cause a shift in our understanding of context. The role of conventional communicative con-texts now seems primarily important as the bearers of the information of how others - the forum one confronts - can be approached and potentially convinced. Relating the context primarily to the forum - instead of to the speaker - can reconcile the idea of contextual restrictions with the simultaneous openness of this context. For example, Noddings, who reconciles herself to the idea that she has to argue to convince others of her positions (1984, p. 6), clearly demonstrates how she fine-tunes her arguments to the - Kantian - forum she addresses. She chooses her arguments, guided by the beliefs of these specific opponents as she observes them, apparently aware of the chances she will not be heard or understood if she arguments otherwise. "But we must realize, also, that one writing on philosophical-educational problems may be handicapped and even rejected in the attempt to bring a new voice to an old domain, particularly when entrance to that domain is gained by uttering the appropriate passwords" (ibid., p. 2).

This means that the relevance of the context with respect to making and justifying knowledge claims should be understood as related to the communicative conventions of those who are to be convinced. If context restricts anything, restrictions should be related to the audience addressed; context restricts the range of arguments this audience might appreciate. This explains why a proponent fine-tunes his argumentation to the forum he has in mind. Stalnaker: "Discourse contexts, I have been suggesting, can be represented by the set of possible situations compatible with the information that is presumed, by the speaker, to be common ground, or information that is shared by all the relevant participants" (1999, p. 101). The restrictive nature of a context as understood here functions as a condition of possibility for - rather than as a hindrance to - convincing others. It makes it possible to choose arguments that are potentially understandable and effective to influence the beliefs and convictions of others. Consequently, this revision of the concept of context results in shifting attention from whatever would become impossible because of it, to what becomes possible thanks to it.

Against this background changing communicative conventions seems more the rule than an exception. Though the acceptability of new claims remains controlled by the norms and concepts that are characteristic of this context, this does not make that context a steely frame. As Brandom formulates it: "For the inferential norms that govern the use of concepts are not handed down to us on tablets from above; they are not guaranteed in advance to be complete or coherent with each other. They are at best constraints that aim us in a direction when assessing novel claims. They neither determine the resultant vector of their interaction, nor are they themselves immune from alteration as a result of the collision of competing claims or inferential commitments that have never before been confronted with one another" (2000, p. 176).

The contextual beliefs, concepts, and norms that regulate the process of mutual conviction are themselves subject to change in the course of this very process; this kind of change is a main raison d'être for communication, and the participants are attuned to this aspect of communicative contexts. This interpretation of context means, that every contribution to communication will change contextual communicative conventions as soon as it is accepted. "Every claim and inference we make at once sustains and transforms the tradition in which the conceptual norms that govern that process are implicit. ... To use a vocabulary is to change it" (ibid., p. 177). Participants register the normative reorientations during the process - that should be considered part and parcel of any communicative practice -, an aspect of participation in communicative practices Brandom (1994, p. xiv) calls score-keeping.

6. The primacy of commitment (II)

The ambition to avoid relativism while maintaining the primacy of commitment results in a conception of commitment that does not prioritize either the feeling-dimension or the content- dimension, and that stresses the willingness to propagate one's own views in the presence of an opposing audience. Relatedly, this ambition results in a conception of context as the changing system of beliefs, concepts and norms that make it possible to express and evaluate new claims In as far as restrictions are involved, they are related to the belief-systems of the audience that do forbid accepting any arbitrary position, but that enable a proponent to formulate potentially convincing arguments in the first place. In other words, they do not function as communicative barriers, but as conditions of the possibility of communication. In this view, restrictive norms do indeed regulate the development and justification of knowledge, but they do not imply the unchangeable rigidity that is characteristic of relativist epistemological views. With respect to the primacy of commitment, the question remains whether it still makes sense to speak of the primacy of commitment, and if so, what does this primacy exactly mean?

In the discussed texts, the primacy of commitment seemed to imply the local - person, group, or discourse-related - and not universal validity of knowledge. However, if communicative contexts are changeable and mutually open, and individual affects are not to be considered as disconnected from content, then the idea of separate communicative contexts should perhaps be abandoned and make place for the idea of one communicative space in which ultimately one form of validity would exist - be it a changeable one. In other words, do the previous analyses of commitment and context disprove the possibility of speaking about local validity? Are we not forced back into the position of the primacy of epistemology? I think not.

To start with, the primacy of epistemology does not only entail the idea of universally true knowledge, but also the idea that such knowledge would be true independent of our psychological or social situation (Williams, 2001, p. 65). And this is not the case, even though we can speak of one coordinating communicative space. On the contrary, as long as we maintain that the justification of knowledge builds on a system of currently accepted claims, a "body of common assumptions" (ibid.), and not on a system of epistemologically privileged foundations, no situation could possibly arise in which knowledge would not depend on psychological or social factors. Consequently, my analyses do not refute antifoundationalism, and the primacy of developing and accepting knowledge remains in the beliefs people endorse in the social and/or the personal realm.

Apart from that, it remains meaningful to speak of local acceptability. Michael Williams points out that all instances of justifying or doubting knowledge - including all attempts to make others change positions - take place in some specific context, i.e. some specific system of assumptions that function as criteria the validity of which is not doubted - at least not for the time being. In order to explain this. Williams discusses the nature of these assumptions in more detail. In his view, we should not simply understand them as replacements for the foundational certainties one has to abandon as an antifoundationalist. In Williams' view, such assumptions do not function as 'foundations'. They do not contribute to 'proving' anything, they rather enable us to ask questions and to formulate answers. For example, we have to assume that the earth did exist five minutes ago if we want to ask and examine historical questions (ibid., p. 160).

Such context-specific assumptions make the existence of a variety of discursive practices possible, and assumptions that - for the time being - function as unquestioned in one context, can be subjected to challenge in another, or in the same at a later point in time. This also explains why it remains a good idea to distinguish communicative contexts from each other, even if they are not considered 'closed' or 'fixed': they give rise to different types of questions, made possible by their specific sets of default assumptions. In this way, the communicative practice in which physical knowledge is developed and tested differs from the communicative practice in which historical knowledge is developed and tested - though in both contexts assumptions change significantly over time as renewed knowledge claims are accepted or rejected.

This pragmatic distinction between communicative contexts does not prevent them from being mutually open. It is possible to move between contexts in various ways. For one, we could imagine 'umbrella-questions' that concern different contexts at the same time. Such as questions that ask for comparisons (Brandom, 2000, P. 171). For example, one could ask for the similarities and the differences in the ways children are studied in psychological and in educationalist contexts. Such questions would imply an 'umbrella-context', that in its turn is characterized by specific constitutive assumptions, such as assumptions about what it means to 'study children'. It is also possible for insights that are developed in one context to cause changes in another context. For example, new conceptions of the physical universe, in which the geocentric picture of the world was abandoned, resulted in a serious undermining of the astrological discursive practice (Williams, 2001, p. 227).

Neither the revised conception of commitment, nor the revised conception of context gives cause to renewed adherence to foundationalism. On the contrary, both rather support the idea of the primacy of commitment with respect to the development and justification of knowledge. The progress of knowledge seems all the more dependent on the claims people want to defend in the presence of specific fora and in the context of specific communicative conventions.

7. Discussion

Not only the position I have called the 'primacy of commitment', but also the way antifoundationalist philosophers of education alerted to it and defended it indicates a specific commitment. In defending their position, these authors also had in mind to make a contribution to reducing social exclusion if only by revealing it. They alerted to the contextual restrictiveness of knowledge claims as a main source of potential exclusion, because these restrictions would keep us from recognizing claims originating from alternative sources or contexts. My revised interpretation of the primacy of commitment makes this view of exclusion no longer tenable. However, the primacy of commitment still has relevant implications with respect to the problem of 'exclusion'. Now the primacy of commitment is related to the willingness to make and defend claims within a specific forum, exclusion gets to mean exclusion from participation in specific discursive practices. This participatory aspect of communication can also be thought of political and ethical importance, which is what Brandom does when he states that: "Our moral worth is our dignity as potential contributors to the conversation. This is what our political institutions have a duty to recognize, secure and promote" (2000, p. 178).

Access to communicative fora can be blocked in various ways. It can be caused by a lack of familiarity with the constitutive assumptions of the involved practice (for example a specific discipline), but refusing access to potential participants also implies participative exclusion. In the latter case, characteristic of the claimer – such as social background or ethnicity – may be used as a criterion for the relevance or the acceptability of the claim. In such situations, social power relations are reflected in the communicative assumptions of this specific practice. History gives ample evidence of this form of participatory exclusion. In most Western countries, women were refused access to the parliamentary-political discourse until far into the twentieth century, and even today some defend the idea that political responsibility should be understood as exclusively male. Fortunately, such assumptions are changeable – at least this one has changed from an unquestionable to a disputable claim.

The antifoundationalist idea of local validity not only inspires philosophers of education, it also worries them. It could result in "intellectual paralysis" (Blake et al., 1998, p. 5), and the tendency to reduce the tasks of philosophy of education to formulating critique. Smeyers (2005) wants to oppose this by suggesting that philosophers of education should make their initial convictions/beliefs – the frame from which they judge things – explicit from the beginning. Making basic beliefs explicit would not annul the limited nature of validity, but it would at least make these limits visible and put each contribution in the right perspective, Smeyers seems to think. Accordingly, as he formulates it, "the...time seems to be ripe for explicit moral (perhaps even political) commitments" (ibid., p. 183). However, it is questionable if and to what degree making initial positions explicit could solve the problem that Smeyers observes. Of course, in as far as it is possible to explicitly formulate any assumptions one makes, doing so would tribute to clarity. However, that would still leave untouched the *discursive* assumptions that regulate how we formulate and defend our claims, including political and moral ones. As discussed above, such discursive assumptions refer to the beliefs, concepts, and criteria of the addressed *forum*, of the current discursive practice in which the protagonist brings his claims to the fore.

Smeyers remarks, however, seem to suggest more than an appeal to philosophers to be clear and hide as few ideas from any audience as possible. He seems to have in mind that people should make explicit the supposedly irreducible *source*, the *basic beliefs* that explain the content of the specific claims they make as protagonists. This only seems a welcome form of openness as long as one assumes that the protagonist has certain 'basic' beliefs that are irreducible and sacrosanct, from which he derives his claims - a state of affairs the addressed forum can only recognise and accept. However, this view of local acceptability that implies the impossibility of bringing locally accepted criteria up for discussion was criticized above, and rejected as relativist. The solution Smeyers suggests holds on to this - relativist - illusion that some 'basic' claims are inevitably exempt from further justification because they are 'basic' for a specific person or group. Holding on to this illusion could result in the idea that making beliefs that are designated as 'basic' explicit, could be a license for refraining from further duties of justification. In my view, the primacy of commitment rather implies the willingness as well as the possibility of defending any claim - explicitly or implicitly made - in the face of doubts, and to change it in the face of convincing counter-arguments. Against this background, there is no reason to fear that we will have to reduce the tasks of philosophy of education to criticizing.

8. References

Audi, R. (2000). *Religious commitment and secular reason.* Cambridge: Cambridge University Press.

Becker, H. S. (1960). Notes on the concept of commitment. *The American Journal of Sociology, 66*(1), 32-40.

Bellah, R.N., Madsen, R., Sullivan, W.M. & Swidler, A. (1985). *Habits of the heart. Individualism and commitment in American life.* Berkeley: University of California Press.

Biesta, G. J. J. (1998). Say You Want a Revolution ... Suggestions for the Impossible Future of Critical Pedagogy. *Educational Theory, 48*(4), 499-510.

Biesta, G.J.J. (1999). Radical intersubjectivity. Reflections on the "different" foundation of education, *Studies in Philosophy and Education, 18*(2), 203-220.

Biesta, G. J. J. (2001). How can philosophy of education be critical? How critical can philosophy of education be? - deconstnictive reflections on children's rights. In F. G. Heyting, D. Lenzen & J. White (Eds.), *Methods in Philosophy of Education* (pp. 123-143). London: Routledge.

Biesta, G. J. J., & Miedema, S. (1999). Dewey in Europe. Ambivalences and contradictions in the turn-of-the-century educational reform. In J. Oelkers & F. Osterwalder (Eds.), *Die neue Erziehung. Beitreige zur Internationaliteit der Reformpaedagogik* (pp. 99-124). Bern: Peter Lang.

Blake, N., Smeyers, P., Smith, R., & Standish, P. (1998). *Thinking again. Education after postmodernism.* London: Bergin & Garvey.

Blustein, J. (1991). *Care and commitment: taking the personal point of view.* New York: Oxford University Press.

Brandom, R.B. (1994). *Making it explicit: Reasoning, representing, and discursive commitment. Cambridge*, MA: Harvard University Press.

Brandom, R.B. (2000). Vocabularies of pragmatism: Synthesizing naturalism and historicism_ in: R.B. Brandom (Ed.), *Rorty and his critics* (pp. 156-183). Malden, MA: Blackwell Publishing.

Buchanan, B. (1974). Building organizational commitment: the socialization of managers in work organizations. *Administrative Science Quarterly, 19*(1). 533-546.

Child, M., Williams, D. D., & Birch, A. J. (1995). Autonomy or heteronomy? Levinas's challenge to modernism and postmodernism. *Educational Theory. 45*(2), 167-189.

Cooper, D. E. (1990). *Existentialism: a reconstruction*. Oxford: Basil Blackwell.

Coser, L. A. (1974). *Greedy institutions: patterns of undivided commitment*. New York: The Free Press.

Craib, I. (1976). *Existentialism and sociology: a study of Jean-Paul Sartre*. London: Cambridge University Press.

De Brabander, R. A. (2003). *Engagement in spiegelschrift. Een confrontatie tussen Maurice Blanchot en Jacques Derrida* (PhD-thesis), Amsterdam: Universiteit van Amsterdam.

De Vries, G. (1992). Bewogen bewegers, in: L. Nauta. et al. (Eds.). *De rol van de intellectueel. Een discussie over distantie en betrokkenheid* (pp. 91-110). Amsterdam: Van Gennep.

Echeverria, E.J. (1981). *Criticism and commitment. Major themes in contemporary 'post- critical' philosophy* (PhD-thesis). Amsterdam: Free University.

Egea-Kuehne, D. (1995). Deconstruction revisited and Derrida's call for academic responsibility. *Educational Theory, 45*(3), 293-309.

Fitzsimons, P., & Smith, G. (2000). Philosophy and indigenous cultural transformation. *Educational Philosophy and Theory, 32*(1), 25-41.

Foucault, M. (1977). *Language, counter-memory, practice: selected essays and interviews* (D. F. Bouchard & S. Simon, Trans.), in D. Bouchard, F. (Ed.). Oxford: Basil Blackwell.

Giroux, H. A. (1997). *Pedagogy and the politics of hope: theory, culture, and schooling: a critical reader*. Boulder: Westview Press.

Gur-Ze'ev, I. (1998). Towards a nonrepressive critical pedagogy, *Educational Theory, 48*(4), 463-86.

Heelas, P. (1996). Emotion talk across cultures, in: R. Heine & P. Gerrod (Eds.), *The emotions. Social, cultural and biological dimensions* (pp. 171-199). London: Sage.

Heyting, F. (2001). Antifoundationalist foundational research - analysing discourse in children's rights to decide. In F. Heyting, D. Lenzen & J. White (Eds.), *Methods in Philosophy and Education* (pp. 108-124). London: Routledge.

Kohli, W. (1998). Critical education and embodied subjects: making the poststructural turn. *Educational Theory, 48*(4), 511-519.

Lieberman, M.S. (1998). *Commitment, value, and moral realism*. Cambridge: Cambridge University Press.

Lovie, S.D. (1992). *Context and commitment. A psychology of science*. London: Harvester Wheatsheaf.

Marshall, J. D. (1996). Doing philosophy of education. In J. D. Marshall (Ed.), *Michel Foucault: Personal Autonomy and Education* (pp. 195-211). Dordrecht: Kluwer Academic Press.

Masschelein, J. (1998). How to imagine something exterior to the system: critical education as problematization. *Educational Theory, 48*(4), 521-530.

Masschelein, J. (2000). Can education still be critical. *Journal of Philosophy of Education, 34*(4), 603-616.

McLaren, P., & Giroux, H. A. (1997). Writing from the margins: geographies of identity, pedagogy, and power. In P. McLaren (Ed.), *Revolutionary Multiculturalism - Pedagogies of Dissent for the New Millenium* (pp. 16-41). Boulder: Westview Press.

Meyer, J.P. & Allen, N.J. (1991). A three-component conceptualization of organizational commitment, *Human Resource Management Review, 1*(1), 61-89.

Montefiore, A. (1975). Part I/Part II, in: A. Montefiore (Ed.), *Neutrality and impartiality. The university and political commitment*. London: Cambridge University Press.

Morgan, S.L. (2005). *On the edge of commitment. Educational attainment and race in the United States*. Stanford: Stanford University Press.

Noddings, N. (1984). *Caring, a feminine approach to ethics and moral education*. Berkeley: University of California Press.

Noddings, N. (1995). Care and moral education, in: W. Kohli (Ed.), *Critical conversations in philosophy of education* (pp. 137-148). New York: Routledge.

Noddings, N. (1998). *Philosophy of education*. Boulder: Westview Press.

Norris, P. (2001). *Digital divide: civic engagement, information poverty, and the internet worldwide*. New York: Cambridge University Press.

Peters, M. (1995a). Philosophy and education: 'after' Wittgenstein. In P. Smeyers & J. D. Marshall (Eds.), *Philosophy and education: Accepting Wittgenstein's challenge* (pp. 189-204). Dordrecht: Kluwer Academic Press.

Peters, M. (1995b). Education and the postmodern condition: Revisiting Jean-Francois Lyotard, *Journal of Philosophy of Education, 29*(3), 387-400.

Peters, M., & Lankshear, C. (1996). Postmodern counternarratives. In H. A. Giroux, C.

Lankshear, P. McLaren & M. Peters (Eds.), *Counternarratives - cultural studies and critical pedagogies in postmodern spaces* (pp. 1-39). New York, etc.: Routledge.

Peters, M., & Marshall, J. D. (1999). *Wittgenstein: philosophy, postmodernism, pedagogy*. Westport: Bergin & Garvey.

Prior McCarthy, L. (1995). Bodies of knowledge. *Studies in Philosophy of Education, 14*, 35-48.

Roth, M. S. (1995). *The ironist's cage. Memory, trauma, and the construction of history.* New York: Colombia University Press.

Rorty, R. (1989). *Contingency, irony, and solidarity.* Cambridge: Cambridge University Press.

Ruhloff, J. (2001). The problematic employment of reason in of philosophy of Bildung and education. In F. Heyting, D. Lenzen & .11 White (Eds.), *Methods in Philosophy of Education* (pp. 57-72). London: Routledge.

Säfström, C. A. (1999). On the way to a postmodern curriculum theory - moving from the question of unity to the question of difference. *Studies in Philosophy of Education, 18*(4), 221-233.

Sartre, J.-P. (1948). Qu'est-ce que la litterature ?, in J.-P. Sartre (Ed.), *Situation.* Paris : Librairie Gallimard.

Sartre, J.-P. (1969). *Being and Nothingness: an essay on phenomenological ontology* (H. E. Barnes, Trans.). London: Routledge.

Scanlan, T. K., Simons, J. P., Carpenter, P. J., Schmidt, G. W., & Keeler, B. (1993). The sport commitment model: measurement development for the youth-sport domain. *Journal of Sport and Exercise Psychology, 15,* 16-38.

Simpson, E. (2000). Knowledge in the postmodern university. *Educational Theory, 50*(2), 157-177.

Smeyers, P. (2005). Idle research, futile theory, and the risk for education: reminders of irony and commitment. *Educational Theory, 55*(2), 165-183.

Snik. G., & Van Haaften, W. (2001). Pedagogisch grondslagenonderzoek. In P. Smeyers & B. Levering (Eds.), *Grondslagen van de wetenschappelijke pedagogiek. Modern en postmodern* (pp. 169-187). Amsterdam: Boom.

Stalnaker, R.C. (1999). *Context and content - Essays on intentionality in speech and thought.* Oxford: Oxford University Press.

Standish, P. (2001). Data return: The sense of the given in educational research, *Journal of Philosophy of Education, 35*(3), 497-518.

Sternberg, R.J. (1987). *The triangle of love: intimacy, passion, commitment.* New York: Basic Books.

Stone, L. (1996). A rhetorical revolution for philosophy of education. *Philosophy of Education Society Yearbook.*

Usher, R., Bryant, I., & Johnston, R. (1997). *Adult Education and the Postmodern Challenge - Learning Beyond the Limits.* London: Routledge.

Van Goor, R. & Heyting, F. (2006). The fruits of irony: Gaining insight into how we make meaning of the world, *Studies in philosophy of education, 25*(6), 479-496.

Van Goor, R., Heyfing, F., & Vreeke, G.-J. (2004). Beyond Foundations. Signs of a New Normativity in Philosophy of Education. *Educational Theory, 54*(2), 173-192.

Weinstein, M. (1995). Social justice, epistemology and educational reform. *Journal of Philosophy of Education, 29*(3), 369-385.

Williams, M. (2001). *Problems of knowledge: a critical introduction to epistemology*. Oxford: Oxford University Press.

Willms, J. D. (2003). *Student engagement at school: a sense of belonging and participation. Results from PISA 2000*. Paris: OECD.

IX
SUMMARIZING AND CONCLUDING REMARKS: COMMITMENT AND ACADEMIC RIGOR IN PHILOSOPHY OF EDUCATION

1. Introduction

The occasion for the research in this thesis was the serious discord between philosophers of education, as regards whatever could still be seen as the central task of the discipline. Although the history of the philosophy of education has appeared to be characterized by the continual questioning and redescription of this task (see chapter one), it seems that as from the late 1990s the issue is more urgent than ever. Against the background of radical doubts that have arisen in respect of the framework that was previously used to legitimize the content of the philosophy of education, hefty debates have taken place on what the discipline might still entail in the future, and even whether it still has a future anyway. Many of those debates have an epistemological background; they revolve around the question of the status that can still be attributed to (educational-philosophical) claims to knowledge. In light of those debates, the content of the philosophy of education is also discussed.

Against the background of the developments that I described, I too question the tasks of a contemporary philosophy of education, and I too take as my point of departure the epistemological doubts that have appeared to be characteristic of contemporary philosophy. In this endeavor I especially focus on the rejection of the foundationalist knowledge model, because it is the rejection of this model by several authoritative philoso-phers that seems to be at the core of the radical doubts within philosophy, and it has clearly left its mark on the philosophy of education. The objective of this thesis has always been to get an idea of what - in this light - might be an acceptable epistemological position in order to subsequently examine what implications such a position has on the content of the philosophy of education, with special attention paid to the role that educational philosophical insight might be able to play in the practice of education and teaching. In this final chapter, I will formulate conclusions on these issues. I will work my way towards those conclusions by first and consecutively summarizing the previous chapters – as from chapter two.

2. Summarizing remarks

The first inquiry of this thesis (chapter two) examined how the rejection of the foundationalist model of justification by some renown philosophers has been received by authors working within the present day field of philosophy of education. Since the practical relevance of philosophy of education - traditionally of utmost importance to the discipline - was, in the past, consistently perceived from a foundationalist perspective, I also investigated as to how the authors of the publications that were analyzed believed practical relevance to have been affected by the rejection of the foundationalist model. Subsequently, I investigated which alternative approaches to justification that these philosophers of education came to adopt, and which consequences they considered these to have for the practical relevance of the discipline.

The main objection to the foundationalist justification model as put forward by the authors examined relates to the epistemological privilege the model awards to either foundational claims or justification procedures. They consider the foremost loss to the practical relevance of philosophy of education to be that of its prescriptive capacity; a capacity wholly dependent - at least according to these authors - on such forms of epistemological privilege. The alternative approaches to justification as put forward by the authors analyzed are all evidence of what I term 'radical contextualism'. In radical contextualism, justification is considered to be a context-bound process that is local and temporary in character.

Accordingly, practical relevance was no longer sought in substantiating recommendations. To the authors examined, practical relevance primarily constituted challenging or, at the very least, revealing the processes of exclusion by calling attention to the restrictive - and thus exclusionary - character of any form of justification. This view on practical relevance, which I have termed 'new normativity', distinguishes itself from the traditional normative philosophy of education in that substantive prescriptions are abandoned and authors confine themselves to the identification of restrictions.

This confinement, however, displays a kind of normativity - and thus prescriptivism - in itself. Ultimately, the suggested approach to philosophy of education displays a - more or less implicit - commitment with 'inclusion'. In this respect, it is important to note, however, that the shared intention of these authors to counteract exclusion does not, here, constitute a new and ultimate, taken for granted, ground for justification - in other words, a new epistemological foundation - for other claims. It is better understood as an attitude; not one that derives its status from any epistemological characteristic, but rather a response to the practical consequences of awarding primary importance to flawed epistemology; namely the fostering of various forms of exclusion. To put it differently, these authors have developed views on justification in which the primacy of epistemology is replaced by a basic practical commitment.

The antifoundationalist authors are not the only philosophers who

154

recognise the uncertainty of knowledge claims that seem to be characteristic of contemporary epistemology. There is quite a large group of philosophers of education that also appreciates the inevitable fallibility of knowledge, but that deals with it in a different, more moderate, way. In order to go about my search for an acceptable epistemological approach in a well-considered way, it is important to also scrutinize the position of those authors. In chapter three, I show that the so-called 'fallibilism' in philosophy of education also rejects the 'old' idea of self-justified foundations as a ultimate ground for the justification of knowledge, thus sharing the idea that knowledge always comes with a certain degree of uncertainty. However, fallibilism upholds that certain beliefs may still be classified as 'more certain' than others in an epistemological sense, and in that sense may serve as justificatory grounds that may be fallible, but that can still be regarded as generally valid.

Referring to two examples of fallibilism in the philosophy of education, I show that fallibilism has two powerful characteristics. Due to its recognition of the fallibility of basic beliefs, fallibilism will continue to critically examine its basic beliefs, and adjust them where necessary. Furthermore, it enables us to forcefully defend theoretical claims, so that it can also make content-specific contributions to educational debates – something that antifoundationalist philosophy of education seems to lack. However, I furthermore argue that fallibilism also raises questions that need further answering. On the one hand, it appears that the epistemological privilege that is still being granted to justificatory grounds, in a philosophical sense cannot be properly defended when abandoning the notion of justificatory grounds that are justified *in themselves*. On the other hand, it appears that the notion of fallible justificatory grounds does not reconcile with the - still advocated - idea of general validity, so that the attribution of general validity might even be evidence of ethnocentrism rather than epistemological certainty.

The questions raised by fallibilism prompted me to look further for an epistemological approach that might be able to find its way around these issues. Considering the fact that the problem is located in the notion of privileged justificatory grounds - thus in the vertical structure of knowledge claims - it seems useful to examine in more detail the contextualist approach of knowledge, in which justification is considered to be embedded in a horizontal network of mutually related beliefs, as proposed by the antifoundationalist authors discussed in chapter two. Such an approach does not take the idea of general validity as its starting point and, in any case, has done away with the concept of granting epistemological privilege. However, the question is whether contextualism does not lumber us with a much greater problem, namely 'relativism'.

The fourth chapter links up with the foregoing on two points. Firstly, it starts from the epistemic uncertainty that characterizes contemporary philosophy of education and the idea of fundamental indecision that is connected with it – by, for instance, the antifoundationalist authors in chapter two. I look into the question of how, in the philosophy of education, we can deal with such indecision in a constructive, but also philosophically acceptable, way. I focus especially on irony as a characteristic instrument that has been used by philosophers in the past when faced with uncertainty or indecision. The central questions are: what may be expected from the use of irony as a philosophical instrument in relation with the indecision that we face in the philosophy of education, and which practical relevance of the philosophy of education may be connected with it? The chapter provides a further link with the previous, as it pays extensive attention to the notion of horizontally ordered justification contexts that is taken into account in Rorty's notion of irony. As such, chapter four also renders insights that are important to my ongoing epistemological search.

In general, it becomes clear in chapter four that 'ironic' philosophers, among whom are Schlegel and Kierkegaard, argue that irony sooner raises questions than it provides answers. It is precisely in the questions evoked by irony, ironic authors believe, that the insights lay, which are generated through its use. On the basis of two recent approaches to irony - Bransen's and Rorty's -, I develop my own interpretation whereby irony generates an insight into the irreducible interplay between the informative content of a claim and the presupposed communicative context in which that claim is made. Of course we must be aware of turning irony, or doubt in general, into a 'pseudo-foundation'. This is accomplished by referring to meta-ironic reflection, which demonstrates that this approach to irony - as other approaches - can, in its turn, only be understood as informative against the background of a presupposition (as part of a presupposed context) about the way in which human meaning-making is accomplished. The analysis resulted in a tool for philosophy of education whereby claims are analyzed as amendments to communicative contexts as these are presupposed by speakers or authors.

This is illustrated by the analyses of scientific-educational debates over the issue of 'students at risk'; the problem of groups of pupils in danger of structurally falling behind in terms of academic achievements - in Dutch also referred to as the 'issue of educational deprivation'. Although the 'actual' communicative context in which an utterance was made can never be reconstructed with any certitude - and, therefore, neither the 'actual' informative content of the claim - such an analysis does provide an insight into the interplay of contextual (for instance, a presupposed audience's views on 'students at risk') and informative (suggestions for defining, or dealing with, 'students at risk') dimensions of claims and their acceptability.

In chapter five, I continue my exploration of contextualism as a possibly acceptable epistemological position that takes account of the inevitable uncertainty of knowledge claims. My first question is how the idea of contexts of justification can be concretized. Firstly, I concentrate on Rorty's idea of vocabularies as justification contexts, which came to the fore in the discussion of his concept of irony in chapter four. It becomes clear that Rorty's interpretation of communicative justification context, on the one hand, offers us useful elements, since - among other things - it shows us how the justification, evaluation, and modification of our ways of speaking can be understood, irrespective of how the world *is*. At the same time, however, Rorty's discursive approach of contextual justification does not completely escape the suggestion of relativism, or arbitrariness, because he does not properly clarify how the transformation, or renewal of justification contexts may be understood.

Next, a dynamic-discursive concept of context is derived from chapter four, which enables me to explain how speakers in the communicative process are, on the one hand, bound by contexts of justification, but at the same time transform, and thus transcend, these contexts - immediately doing away with the reproach of relativism. Thinking in terms of the idea of a dynamic-discursive context, an image emerges of justification that, in principle, corresponds to the way in which we deal with justification in our day-to-day conversation, but of which I argue that it can also be used for developing a more academic-intellectually oriented epistemology. I thus arrived at a beginning of an epistemology that I call 'discursive', as the justification and development of knowledge is made fully dependent on the ongoing exchange of claims, objections, and arguments between the participants in a communicative process.

In relation to the possible consequences of such a discursive epistemology for a stance on the tasks of the philosophy of education, in chapter five I conclude that there is no longer any direct epistemological reason for any prescriptive role of the philosophy of education whatsoever – as was, and still is, argued by fallibilist philosophers of education. Then again, the plea for a philosophy of education that limits itself to clarifying contextual restrictions - as submitted by the antifoundationalist philosophers of education in chapter two - does not seem to offer a solution either: firstly, because it itself is also normative, so that, at the least, it radiates the suggestion of prescriptiveness; and, secondly, because it, therefore, does not seem to be contextualist in its actual sense. The notion of restrictions that are inevitably attached to justification, thus rather acts as a pseudo-foundation for the plea for a specific interpretation of the philosophy of education. I believe that the proposed model, therefore, does not optimally use the possibilities offered by a contextual approach towards the philosophy of education.

Chapter six offers a further elaboration of the notion of dynamic-discursive contexts of justification, focusing primarily on the activity of the individual participant in the communication. That focus is needed to investigate whether a discursive epistemology does not inevitably imply conventionalism. This is investigated in light of the question of what it means for a subject to learn to use a language. In view of the potentially practical relevance of the philosophy of education, it is also examined whether answering this question can help us form an idea of how best to deal with 'differential academic language proficiency', which is a topic that is related to the issue of 'students at risk' that is addressed in chapter four.

I argue that the learning of a language can best be understood as an active participation in a process of practice-based communication (negotiation) about how the world *is*, resulting in a conceptualization of the world: a process denoted as 'world-making'. Differences in '(academic) language proficiency' should not, therefore, be construed as different levels of skill in the use of a linguistic apparatus but rather as different levels of expertise acquired through participating in this process of 'world-making'. Approached from this perspective, one may argue for awarding a more substantial role to cooperative problem-solving - which involves the (re-)construction of meaning - in dealing with 'differential language proficiency' in schools than has been done so far.

Based on the insights gained in chapter six, the notion of a discursive epistemological position is further developed in chapter seven. I make it clear as to how communicative processes may be regarded as practice-based negotiation on how the world is. In that negotiation, participants in the communication play the role of both the initiator of the transformation of the practice-based presuppositions on how the world works and the conservator of – another part of – that collection of presuppositions. This picture immediately shows that this approach does not allow for conservative conventionalism.

In such an approach, the development of knowledge appears as a critical-discursive process in which even the most fundamental presuppositions can be put up for discussion when there seems reason to do so within the communication. Then again, certain presuppositions may last a very long time when they satisfy. In such cases one may speak of a state of 'reflective equilibrium'. However, even in such a state of – relative – equilibrium the justification context is continuously evaluated, added to, and modified, and the possibility always remains that the equilibrium may be abruptly upset whenever some of the, what we thought, most basic presuppositions are put up for discussion successfully.

In relation to the issue of the possibility of scientific progress within such an approach to knowledge, it is further submitted that such a progress can only be spoken of in light of specific criteria for progress within a specific scientific communicative practice, where those criteria themselves are also simply a part of the ongoing scientific debate and, therefore, may be put up for discussion too. Just like in all other forms of communication, contributions to the scientific discourse always entail attunement to the common presuppositions, on the one hand, but at the same time a proposal to revise some of those presuppositions, on the other hand. In line with the notion of expertise as developed in chapter six, I subsequently suggest that academic expertise then might best be understood as the ability to excel in the conducting of research according to the common criteria for academic inquiry, as well as the capability to critically, and successfully, bring up for discussion even those beliefs that were deemed to be basic.

I, therefore, argue that the epistemic criteria for what may apply as an acceptable knowledge contribution to educational-philosophical debates are not set before that debate either, but are rather a part of it – which makes them dependent on whatever the participants in the communication deem important. I subsequently submit that, in that respect, we may indeed – as tentatively suggested in chapter two - refer to a shift in the philosophy of education from a 'primacy of epistemology' to a 'primacy of commitment'. Since it has not become clear in the previous chapters as to how such an idea of a so-called 'primacy of commitment' might actually be properly understood, in chapter eight I examine what that idea might entail and what the consequences of such an idea might be for the philosophy of education, with that special attention is paid to its practical relevance.

Drawing on the previously gained insights concerning communicative contexts, in chapter eight I argue that the 'primacy of commitment' in philosophy of education may eventually be best understood as the notion that the development and justification of educational-philosophical knowledge can ultimately be reduced to a process wherein a proponent makes a claim and is prepared to defend that claim in front of a specific audience - as perceived or imagined by the proponent. It is precisely in the willingness of the proponent to present and defend this specific claim in front of this specific audience, where the commitment should be located.

3. Discursive epistemology

Now that the conclusions in the previous chapters have been summarized, the general questions that were formulated at the beginning of this thesis, and that were addressed in different places throughout this thesis, can be answered. This firstly concerns the issue of an epistemological approach that is able to deal with the inevitable fallibility of knowledge claims in an acceptable way. The central question is when one can speak of knowledge, a question that I have interpreted in terms of 'epistemic entitlement'. It then concerns the question when someone can be considered justified to make a certain claim - concerning how the world is. Within the discursive epistemological position that I defend here, one can speak of 'epistemic entitlement' when a claim of a speaker within specific communicative practices is accepted as valid by the audience. Such a situation will only exist if the participants in the communicative practice – the audience – believe that the speaker has sufficient reasons to make that claim. If doubts exist, the speaker will be asked to further support his claim with additional reasons. When the speaker is not able to match his communicative contribution to some minimal epistemic standards that presently apply within the communicative practice at hand, the claim will be rejected and will not become part of the communicative context. In the case that a speaker is deemed justified to make a claim – in other words, when 'epistemic entitlement' applies – the claim made will become part of the collection of presuppositions that are considered to be shared within the communicative practice and to which the participants in the communication may refer to in the further process of the communication as 'self-evident', at least for the time being. This 'presumed common ground', or the communicative context, thus forms the 'body of knowledge' as it were, which can continually be appealed to in the communicative process. In principle, each element of this 'body of knowledge' may be put up for discussion at any moment in the communication. However, since the communicative context is at the same time the necessary condition for the possibility to do so, not all presuppositions that are part of it can be put up for discussion at the same time without breaking off the communication altogether.

This makes it clear that, within a discursive epistemological approach, knowledge is not regarded as a relatively constant collection of propositions that are deemed to guarantee a certain certainty about how the world is, irrespective of a certain communicative practice. Knowledge is, in any case, regarded as bound by a specific communicative practice. However, the 'body of knowledge' is not determined by the communicative practice within which it is generated and applied either, since there are no unshakeable restrictions binding justification within communicative practices. The presupposed knowledge-base within a communicative practice is both the result and part of the ongoing communicative process of making and defending claims, within which, if necessary, additional reasons are requested and provided - whether, or not, to everyone's satisfaction. The fact that knowledge of the world is thus viewed as a provisional outcome of an ongoing communicative negotiation process reveals the social-constructivist nature of a discursive-epistemological approach to knowledge.

Such an approach to knowledge also has consequences for our understanding of academic inquiry and academic knowledge. Among other things, it makes clear that science has no exclusive, or in any way privileged, access to knowledge about the world. After all, science, or the scientific discourse is merely one possible communicative practice within which knowledge claims are generated, applied, and discussed. This conclusion, however, does not imply that no special status can be attributed to scientific knowledge claims. Science may still be regarded as an authority when a *specific* form of knowledge is concerned, one which can be characterized with terms such as 'validity', 'reliability', or 'verifiability', etc. - even if, or actually because – their meaning is continually renewed over time. Whether science, or scientific knowledge, is regarded an authority within a society, rests – just like in the case of (insights from) other communicative practices, such as religion or art – on whatever science yields in the eyes of the recipients, and this social appreciation may also vary historically.

Within the academic world, each discipline may be seen as a separate communicative (sub)practice, each dealing its own questions, concepts, and methods. The same goes for the philosophy of education. In that sense, within a discursive epistemology there is room for educational-philosophical knowledge claims with their own integrity and with their own – philosophical – acceptability criteria. However, insofar as specific educational-philosophical knowledge exists, it also applies that it is both an outcome and a part of a specific – educational philosophical – discourse, so that whatever may be regarded as academically acceptable educational-philosophical knowledge will always be part of an ongoing negotiation process.

4. Tasks and possibilities of philosophy of education

Now it has been clarified as to how educational-philosophical knowledge can be understood from a discursive epistemological standpoint, we can raise the question of which consequences may be attached to that position when it comes to the tasks and possibilities of philosophy of education.

In the first place, I found that there seems to be no apparent epistemological reason for philosophy of education to limit itself to criticism, as in bringing to attention the contextual restrictions that restrain justification. In chapter four, however, I show that reconstructing elements of the presupposed communicative contexts of justification can be valuable. The illustration of my interpretation of irony as a philosophical tool showed, among other things, which taken-for-granted presuppositions may possibly lie hidden behind influential publications, such as those by Slavin and Maddin about dealing with the problem of so-called 'students at risk'. These publications might well be attuned to an audience that presupposedly blindly accepts the idea that education primarily revolves around the acquisition of language and cognitive skills, or the idea that, at all costs, we should prevent certain children in education from running the risk of lagging behind in terms of school results. Such a clarification of these ideas immediately raises the question of whether they are really as self-evident as they might be presupposed within the analyzed communicative practices, which may be regarded as a success for the critical-reflective capability of the philosophy of education. Besides for the critical-reflective gain that this ironic analysis might render, the ironic analysis has furthermore generated insight into how the construction of meaning could (!) be understood on the basis of the exchange between the content of a claim and the communicative context within which that claim has been made. For that matter, my critical-reflective contribution to the philosophy of education can itself also be seen as a theoretical and content-specific contribution to the communication, which would imply that it can also only be understood in light of a reconstruction of the presuppositions that the speaker ascribes to their audience.

Viewed from the perspective of a discursive epistemology, each educational-philosophical contribution is itself thus inevitably both a theoretical contribution - since it is an amendment to the actual communicative context - and a critical contribution - since a part of the actual communicative context is always put up for discussion. Chapter six shows a second example of a theoretical and content-specific contribution to the educational-philosophical discourse. In this chapter, I make a substantiated proposal to understand a subject's learning to use a language in terms of an active participation in an ongoing practice-based negotiation on how the world is. This immediately shows that a philosophy of education that rejects epistemological privilege is perfectly capable of forcefully propagating and/or defending ideas. However, understood from a discursive epistemology, the acceptance of the claims, or the persuasive power of the arguments, can only be understood in light of the - perhaps occasionally adjusted - acceptability standards as applied by the audience in question - in this case an educational-philosophical audience, which

162

immediately eliminates any possibility of being able to establish epistemic certainty or general validity. The fact that this chapter - as were chapters two and four – were published in a peer-reviewed educational philosophical journal indicates that the claims and arguments that were put forward were found acceptable by at least a (small) part of the educational-philosophical audience.

It thus appears that a discursive epistemology does not set any direct conditions or limits as to what the philosophy of education should concern itself with, which confirms that the so-called 'primacy of epistemology' is done away with. I have tried to show that epistemological questions are indeed relevant to educational philosophers, but that they do not precede the query into the content of the discipline. Just like that query, they are simply part of the ongoing negotiation on how the world - of the philosophy of education - works. However, the query into the content of the philosophy of education and epistemological questions are not inseparable, as Cooper, for instance, seems to think (1998, p. 212). In light of the claim that knowledge claims have local validity, however, it is quite plausible to urge educational philosophers to exercise restraint when it comes to linking pretenses to their educational philosophical insights beyond the 'boundaries' of their own communicative practice - practical pretenses for instance.

Now that I have argued that the set of instruments available to philosophy of education is not limited beforehand, but that at minimum it is open to discussion as to which methods educational philosophers can use, it seems relevant to examine which conclusions may be drawn in relation to the subjects or topics to which the philosophy of education should, or could, apply itself. In line with the conclusion in chapter seven, it seems plausible that philosophers of education, like any other academic, should simply occupy themselves with conducting content-specific research in accordance with the common standards and methodological requirements for proper research. In this thesis I show that this research may be focused on different areas by concentrating on the philosophy of education as an academic discipline (chapters one, two, four, and eight), by taking a more empirical-scientific discussion - on how to deal with 'students at risk' - as a starting point for my illustrating analyses in chapter four, and by examining whether certain philosophical insight could lead to another (philosophical) understanding of the educational practice as regards to dealing with 'differential academic language proficiency' in chapter six. Chapter seven, however, also suggests that scientists should also assign themselves the task to critically question presuppositions that are deemed basic within their own discipline - in respect of the how, what, and why of discipline-based research activities, among other things. It should be clear by now that this task has been focused on throughout this entire thesis, in the shape of a discussion of other educational philosophical positions on this subject (again, see chapters one, two, four, and eight), as well as in the development of my own position on the task and possibilities of philosophy of education.

A more content-related theme that I addressed at different times in this thesis was 'fighting exclusion'. Chapter two shows how this fight was the reason for antifoundationalist philosophers of education to call for a philosophy of education that limits itself to drawing attention to the constraints attached to any justification of educational claims. That way, the structural excluding effect of a justification would come into view, creating room for alternative contributions that could not have been envisaged before: alternative meanings, different 'embodied' personal viewpoints, or contributions from incommensurable discourses. Chapters four and six confronted us with educational debates that pay attention to attempts to ensure that certain groups of pupils will in future no longer structurally fall by the wayside where their school results are concerned. In view of these contributions, the desirability of fighting exclusion seems to be presupposed as taken-for-granted within the wider educational communicative practice.

Next, in chapter eight, I make a proposal for a certain approach towards 'fighting exclusion'. I argue that in an approach in which the primacy of commitment is related to the willingness to make, and defend, claims within a certain forum, exclusion might be best understood in terms of being excluded from participating in specific communicative practices. The exclusion may come about by applying acceptability standards within a certain communicative practice, which cannot be met due to a lack of communicative expertise, or simply because you are a certain person, belong to a certain group, or have certain characteristics. In general, I suggest that fighting against exclusion directly or indirectly always has to do with the putting up for discussion practice-based acceptability norms that are responsible for a certain exclusion. In relation to this, I propose in chapter six that in the attempts to let children not fall by the wayside when it comes to their development of academic language proficiency, it might be desirable to acknowledge children as full, active participants within the communicative practice in schools more and at an earlier stage, and to further exploit that aspect in the design of the curriculum.

An implication of my position, however, is that the fighting exclusion itself, as a standard, may also be put up for discussion. For instance, at several moments in this thesis I very clearly show myself in favor of applying quite strict standards for academic inquiry in general

164

and in the philosophy of education in particular, and to also rigidly see to it that such standards are followed. This will entail a clear exclusion of certain contributions to the communication, unless it is able to successfully transform one or more of the standards prevailing at that moment.

With each task that I set for myself in this thesis, and at each moment that I explicitly addressed the tasks of the philosophy of education, as a speaker I wanted to contribute to the academic educational-philosophical discourse. Given my idea of the primacy of commitment, it is my own commitment that came to the fore. It is my own decision to make this contribution and to do so in this manner, with an eye to transforming – however small these changes may be – the educational-philosophical discourse. This reveals a subjective element in the way the educational philosophical communicative practice – like any other communicative practice - takes shape. The separate contributions to the communication may be biased by personal taste, moral disapproval, instinct, political conviction, philosophical preferences, etc., which will also leave their mark in the communicative practice as a whole. Besides for the epistemic or conventional ones, personal (existential, moral, political, and aesthetic) elements will thus also play a role in the evolution of communicative practices - including academic practices. This in no way means that communicative contributions, such as this thesis, could be reduced to being a product of subjective expression. After all, the commitment I refer to here is not subjective, but primarily communicative. It only finds expression in the concepts that are used and the methods that are elected to attune to a certain audience on the grounds of an - informed - assessment of the acceptability standards that are shared by that audience. Speakers are inevitably bound by the presuppositions deemed to be shared by their audiences; this is the case for the formation of their ideas, but certainly also for the expression and defense of those ideas. Here, the primacy of commitment ultimately means that I am driven to bring to the fore whatever I bring to the fore in reaction and attunement to the educational-philosophical forum, trying to do justice to that forum, but also wanting to transform it by means of my contribution, and that I feel the drive and responsibility to defend my contribution within that forum whenever there is doubt that I have sufficient reasons to make my claims, in other words: when my 'epistemic entitlement' is put up for discussion.

In light of these conclusions it becomes clear that a primacy of commitment in the philosophy of education does not have to erode the discipline's academic rigor. On the contrary, my position shows that if academic disciplines wish to keep fulfilling their roles as critical-reflective knowledge authorities, there is a clear task and responsibility for each individual academic to see to it that the prevailing criteria for academic endeavor are met, and that the tenability of those standards themselves are periodically assessed - especially in light of the question of whether, or not, a status of untouchability is unjustly ascribed to those very same criteria, be it intended or unintended. Thus, commitment does not seem to be a threat to 'academic rigor', but rather a part of the process in which the value of this rigor, including the criteria that it implies, must be realized time and time again.

5. The possible contribution of the philosophy of education to practical-educational discourse

Finally, I wish to address the consequences of the foregoing for an idea about the practical relevance of the philosophy of education, a topic that of old has played a key role in the philosophy of education (see chapter one). Whereas within the educational-philosophical discourse the acceptability of a communicative contribution, for instance, will be more dependent on the extent to which the prevailing academic standards are met, within the practical-educational discourse the expected yield for solving a practical educational problem will be more decisive. This does, however, not imply that a mutual influence between the separable discourses is impossible. Contribution from the one discourse will certainly exert influence on the other because there may be 'umbrella questions', for instance. In both discourses, the goals of education may be questioned, for example. This only shows that a contribution from one communicative practice within the other will emerge in light of a different communicative context, so that it will also be attributed a different meaning. Whether within a discourse a contribution will be seen as relevant, may exert influence, and what that influence might possible be can, therefore, not be assessed within the communicative context of another discourse.

Hence, however, the potential interchange between philosophy of education and educational practice may take shape, it will assuredly not proceed according to any established pattern. Consequently, it will be impossible for philosophy of education to control its practical relevance or impact. Even if philosophy of education is to have any practical relevance, this would more likely be 'inspirational' rather than 'derivational'. Against that background, the notion of a philosophy of education that justifies its existence by referring to the role it might play in educational practice appears to be, at the very least, problematic. Nevertheless, the question of the practical relevance of philosophy of education has not yet been resolved. For now, as seen from the perspective set out above, the matter is simply that philosophy of education does not need to appeal to its practical relevance in order to formulate its central tasks. After all, if a philosophical-educational discourse can be regarded as a distinct communicative practice, with its own central tasks, questions, concepts, methods of inquiry, and acceptability norms then this discourse has its own integrity, irrespective of other communicative practices. How, exactly, philosophy of education may perceive its own central task, or the way to fulfill such a task remains under discussion. It does not seem judicious to allow the content of philosophy of education to be determined by its potential impact on educational practice because this impact, if it even were to occur, cannot be predicted. For a philosopher of education to do so would be like an author to consider the actual writing of a book worthless unless he or she were assured in advance that readers would appreciate the work on his/her personal motivation for undertaking the endeavor. To do so would be to undervalue the specific integrity of the discipline and to overrate its capacity to control if and how its insights may be recognized and reproduced.

There are many ways of responding to the preceding conclusion. Some philosophers of education try to help their 'inspirational' impact along by focusing on questions (they believe to be) posed by educational practice. An example is the Philosophy of Education Society Great Britain issuing a series of publications - of which the title, 'IMPACT', is perhaps indicative of the organization's desired role in the practical debate - in which British educational policy is critically examined from a philoso-phical perspective (see f.i. Curren, 2009). The same happened in the Netherlands two decades ago when a number of eminent Dutch philoso-phers of education responded to the then minister of education's proposal that schools should take their role in upbringing more seriously (Ritzen, 1992)[9]. Analysis of such contributions of philosophers of education reveals an interesting difference between philosophical educational re-search and empirical educational research. Even though the efforts can be regarded as valuable - if only to show that philosophy is capable of dealing with real-life educational issues -, one may wonder what the actual impact that such publications on educational practice or educational policy-making has been. Over time, philosophical-educational research findings have come to meet significantly less recognition than have the outcomes of empirical-educational studies. In view of the preceding, this must be related to the role of acceptability norms in different com-municative contexts. Evidently, the fact that something has been studied empirically has become incorporated as an acceptability norm in the

---

9   They did so in a special issue of *'Pedagogisch tijdschrift. Forum voor opvoedkunde'* (1992, Vol. 17, issue 2).

practical educational discourse - and most certainly so in communication about educational policy. This has led to a transformation in empirical-educational scholarship. The employment of 'empirical' as an important criterion by educational practice has induced academic researchers who aim to be practically relevant to increasingly see the making of 'evidence based' recommendations as their 'core business'. In this interplay between educational practice and educational science, it seems to be acceptable that many, previously current, practical educational insights that, by nature, lean towards the 'philosophical' simply disappear. The emphasis in the publications by both Slavin & Madden and Schweinhardt & Weikart on the quantitatively established effectivity of the programs they tested - and developed - for dealing with 'students at risk', for example, appears to mark a communicative process in which such research findings are the deciding factor in the choice of educational programs (see chapter four). Consequently, practical considerations that had previously played a central role - for instance, considerations relating to the aims of education; the school-readiness of children; or schools' ideological backgrounds - seem to fade away. In that sense, the far-reaching impact of empirical-pedagogical research on educational practice may constitute an enrichment in some respects, but definitely also an impoverishment in others.

6. References

Cooper, D. E. (1998). Interpretation, construction and the 'postmodern' ethos, in: David Carr (ed.), *Education, knowledge and truth: beyond the postmodern impasse* (pp. 37-49). London: Routledge.

Curren, R. R. (Ed.)(2009). *Education for sustainable development: a philosophical assessment.* London: Philosophy of Education Society of Great Britain.

Ritzen, J. M. M. (1992). *De pedagogische opdracht van het onderwijs: een uitnodiging tot gezamenlijke actie.* Zoetermeer: Ministerie van Onderwijs, Cultuur & Wetenschappen

.

NEDERLANDSE SAMENVATTING (DUTCH SUMMARY)

OMGAAN MET EPISTEMISCHE ONZEKERHEID: KENMERKEN, MOGELIJKHEDEN EN BEPERKINGEN VAN EEN DISCURSIEF-CONTEXTUALISTISCHE BENADERING VAN DE OPVOEDINGSFILOSOFIE

I. Inleiding: een korte reconstructie van de ontwikkeling van de opvoedingsfilosofie als academische discipline.

Twijfels over de 'grote verhalen' die van oudsher werden ingezet om de taken en werkwijzen van de opvoedingsfilosofie te legitimeren, hebben ook de discussie doen oplaaien over de mogelijke bijdrage die dit vakgebied zou kunnen leveren aan het pedagogisch denken. Hoe de onzekerheid die met deze twijfels samenhangt kan worden begrepen en welke consequenties hieraan kunnen worden verbonden voor de taken en mogelijkheden van de opvoedingsfilosofie is het centrale onderwerp van dit proefschrift. Omdat de discussie hieromtrent binnen de opvoedingsfilosofie voor een belangrijk deel gevoerd wordt op grond van epistemologische overwegingen, zal ook de beantwoording van de gestelde vragen voor een belangrijk deel plaatsvinden tegen de achtergrond van de receptie van recente epistemologische inzichten binnen de opvoedingsfilosofie.

In het eerste hoofdstuk ga ik in op de ontwikkeling van de opvoedingsfilosofie als academische discipline. Het wordt duidelijk dat het bevragen van haar eigen taakstelling niet nieuw is voor de opvoedingsfilosofie, maar dat het haar al typeert vanaf het moment dat ze zich vormde tot een onafhankelijk vakgebied. Beginnend vanuit een voornamelijk levensbeschouwelijke oriëntatie ontwikkelde de opvoedingsfilosofie zich in de loop van de twintigste eeuw, veelal in lijn met bredere bewegingen binnen de (wetenschaps)filosofie, tot een eigen academische discipline. Hoofdstuk één behandelt verschillende stromingen en perspectieven die in de loop van de tijd binnen deze discipline ontstonden, elk met een eigen idee over hoe invulling gegeven zou moeten worden aan de opvoedingsfilosofie en wat dat zou moeten opleveren voor het pedagogisch denken.

De opvoedingsfilosofie kenmerkte zich dus altijd al door discussies over haar invulling. Aan het einde van de twintigste eeuw leek het debat echter een meer radicale wending te nemen. Een steeds grotere groep opvoedingsfilosofen omarmde het idee dat er eigenlijk in algemene zin helemaal niets te zeggen valt over de richting waarin de opvoedings-filosofie zich zou moeten ontwikkelen, waardoor de opvoedingsfilosofie zich maar beter neer zou kunnen leggen bij het idee dat we overgeleverd zijn aan een onverenigbare pluraliteit aan mogelijke perspectieven. Hoewel dit

169

idee van meet af aan ook fel bestreden is door andere opvoedingsfilosofen, hebben de radicale twijfels aan de mogelijkheid van een over-
koepelende benadering die werden geopperd wel de vraag opgeroepen wat
er nog verwacht kan worden van een toekomstige opvoedingsfilosofie.

Tegen deze achtergrond stel ik drie centrale vragen die als een
rode draad door het proefschrift heenlopen. Hoe kunnen we de – voornamelijk epistemologische – kwesties die aan de basis lijken te liggen van
de recente discussies over de invulling van de opvoedingsfilosofie begrij-
pen? Welke consequenties voor de taakstelling van de opvoedingsfilosofie kunnen verbonden worden met de ontwikkelde inzichten? En wat
betekent dit alles voor de mogelijke bijdrage die de opvoedingsfilosofie
kan leveren aan het pedagogisch denken? Hoewel het proefschrift gezien
kan worden als een doorlopend verhaal waarin wordt gezocht naar
antwoorden op bovengenoemde vragen, zijn hoofdstukken twee, vier, zes
en acht ook zelfstandig te lezen, omdat ze zijn geschreven als op zichzelf
staande onderzoeksartikelen.

II. Voorbij fundamenten – tekenen van een nieuwe normativiteit in de opvoedingsfilosofie (gepubliceerd in: Educational Theory, 54(2), 173-192)

In hoofdstuk twee betoog ik dat de onzekerheid die een gedeelte van de
recente opvoedingsfilosofie kenmerkt voor de belangrijk deel voortkomt
uit de afwijzing van het zogenaamde fundationalistische model voor de
rechtvaardiging van kennisaanspraken. Ik onderzoek voorts hoe de afwijzing van het fundationalistische rechtvaardigingsmodel is ontvangen
binnen opvoedingsfilosofische publicaties. Omdat de praktische relevantie
van de opvoedingsfilosofie – van oudsher zeer belangrijk voor deze
discipline – in het verleden veelal begrepen werd vanuit een fundatio-
nalistisch perspectief, onderzoek ik ook hoe die praktische relevantie
volgens de auteurs van de onderzochte publicaties met de afwijzing van
het fundationalistische model is aangetast. Vervolgens ga ik na welke
alternatieve benaderingen van rechtvaardiging van uitspraken deze
opvoedingsfilosofen dan hanteren en welke consequenties zij daaraan
verbinden voor de praktische relevantie van het vak.

Het belangrijkste bezwaar tegen het fundationalistische rechtvaardigingsmodel van de groep onderzochte auteurs blijkt betrekking te
hebben op het epistemologisch privilege dat binnen dit model wordt
toegekend aan rechtvaardigingsgronden, dan wel rechtvaardigings-
procedures. Als belangrijkste verlies voor de praktische relevantie van de
opvoedingsfilosofie wordt daarbij met name een verlies aan prescriptief
vermogen aangemerkt, een vermogen dat – althans volgens de onderzochte auteurs – blijkt te staan of vallen met de genoemde vormen van

epistemologisch privilege. De alternatieve benaderingen van rechtvaardiging die de onderzochte auteurs voorstellen blijken allemaal de vorm te hebben van wat ik een 'radicaal-contextualisme' noem. Rechtvaardiging van beweringen verschijnt daarbinnen als een contextgebonden proces met een lokaal én momentaan karakter.

In lijn hiermee blijken de onderzochte opvoedingsfilosofische auteurs de praktische relevantie van de opvoedingsfilosofie ook niet langer te zoeken in het doen van inhoudelijke aanbevelingen. Zij zien de praktische relevantie van het vak vooral in het bestrijden, of ten minste aan het licht brengen, van processen van uitsluiting, door het beperkende – en daardoor uitsluitende – karakter van pedagogische uitspraken onder de aandacht te brengen. Deze visie op praktische relevantie kwalificeer ik als een 'nieuwe normativiteit', die zich onderscheidt van de traditionele normativiteit binnen de opvoedingsfilosofie doordat men afziet van inhoudelijke prescripties en zich beperkt tot het aanwijzen van contextuele restricties. Uiteindelijk verraadt deze benadering echter mijns inziens toch een – zij het meer impliciet – inhoudelijk engagement met zoiets als 'inclusie'.

Het is in dit verband van belang om op te merken dat de door deze auteurs gedeelde intentie om uitsluiting te bestrijden hier geen dienst doet als nieuwe als vanzelfsprekend gehanteerde uiteindelijke rechtvaardigingsgrond – oftewel als nieuw epistemologisch fundament – voor andere beweringen. Het moet begrepen worden als een attitude die zijn status niet ontleent aan enig epistemologisch kenmerk, maar die wordt voorgesteld als reactie op de praktische consequenties van het primair stellen van de epistemologie, te weten: het in de hand werken van verschillende vormen van uitsluiting. Deze auteurs ontwikkelen anders gezegd visies op de opvoedingsfilosofie waarin het 'primaat van de epistemologie' is vervangen door een 'primaat van engagement', omdat een basaal praktisch engagement richtinggevend wordt bevonden voor de invulling van de opvoedingsfilosofie. Deze reactie op de vermeende tekortkomingen van een fundationalistische opvoedingsfilosofie, lijkt blijk te geven van een vooronderstelde visie van de opvoedingsfilosofie als vakgebied waarvan de praktische relevantie koste wat het kost overeind gehouden dient te worden.

III. Epistemologische inzichten en consequenties voor de opvoedingsfilosofie I: fundationalisme, fallibalisme en contextualisme

De antifundationalistische auteurs die zijn onderzocht in hoofdstuk twee zijn niet de enige opvoedingsfilosofen die het idee onderschrijven dat kennis onvermijdelijk in een bepaalde mate begrensd is, een idee dat kenmerkend lijkt voor de hedendaagse epistemologie. Er is een grote

groep opvoedingsfilosofische auteurs die ook uitgaat van de onover-
komelijke feilbaarheid van kennisaanspraken, maar die er op een andere,
meer gematigde, manier mee omgaat. In mijn poging te komen tot een in
filosofische zin acceptabele interpretatie van het onzekere karakter van
kennisaanspraken voor de opvoedingsfilosofie, is het van belang om ook
de opvattingen van deze auteurs nader onder de loep te nemen. In
hoofdstuk drie laat ik zien dat het zogenaamde 'fallibalisme' in de
opvoedingsfilosofie ook het 'oude' fundationalistische idee van in zichzelf
gerechtvaardigde fundamenten als uiteindelijke grond voor de rechtvaardiging van kennisaanspraken verwerpt, waarmee ze ook het idee omarmen
dat er bij kennis altijd sprake is van een bepaalde mate van onzekerheid.
Het fallibalisme houdt echter – in tegenstelling tot het antifundationalisme
– vast aan het idee dat bepaalde overtuigingen nog altijd aangemerkt
kunnen worden als in epistemologische zin 'meer zeker' dan andere. Zulke
overtuigingen kunnen volgens deze auteurs toch dienen als uiteindelijke
rechtvaardigingsgronden die dan misschien wel feilbaar zijn, maar
waaraan toch een algemene geldigheid kan worden toegekend, omdat het
de best mogelijke gronden betreft die ons op dit moment ter beschikking
staan.

Gebruik makend van twee voorbeelden van het fallibalisme in de
opvoedingsfilosofie – ontleend aan Siegel en Carr – laat ik zien dat het
fallibalisme twee krachtige kenmerken heeft. Door de onderkenning van
de feilbaarheid van die overtuigingen die kunnen dienen als uiteindelijke
rechtvaardigingsgronden zal het fallibalisme deze overtuigingen van tijd
tot tijd aan een nader kritisch onderzoek blijven onderwerpen, en ze indien
nodig vervangen of bijstellen, hetgeen het kritisch-reflectief vermogen
van een fallibalistische opvoedingsfilosofie verhoogt. Daarenboven maakt
het fallibalisme het mogelijk om theoretische opvoedingsfilosofische
claims met kracht te verdedigen, waardoor een specifieke inhoudelijke
bijdrage kan worden geleverd aan pedagogische debatten – een vermogen
dat de antifundationalistische opvoedingsfilosofie lijkt te ontberen. Ik
beargumenteer echter verder dat het fallibalisme ook vragen oproept. Ik
betoog enerzijds dat het epistemologisch privilege dat binnen het
fallibalisme nog altijd wordt toegekend aan rechtvaardigingsgronden in
filosofische zin eigenlijk niet goed verdedigd kan worden. Anderzijds
betoog ik dat het idee van feilbare rechtvaardigingsgronden niet goed
verenigbaar is met de claim op algemene geldigheid die binnen het fallibalisme wordt gedaan, waardoor de toekenning van deze algemene
geldigheid wel eens eerder zou kunnen getuigen van etnocentrisme dan
van epistemologische legitimiteit.

De vragen die het fallibalisme oproept zetten me aan om toch
verder te zoeken naar een epistemologische benadering die mij misschien

beter in staat stelt om op een acceptabele wijze met epistemische onzekerheid om te gaan. Gezien het feit dat de gediagnostiseerde problemen vooral verbonden lijken met de hiërarchische – of verticale – voorstelling van de relatie tussen kennisclaims en rechtaardigingsgronden lijkt het zinvol om een contextualistische benadering, waarin kennis en rechtvaardiging worden beschouwd als verankerd in een horizontaal geordend netwerk van gelijkwaardige, onderling gerelateerde overtuigingen – zoals gehanteerd door veel antifundationalistische opvoedingsfilosofen –, nader op haar merites te onderzoeken. Zo'n contextualistische benadering ziet in elk geval af van het toekennen van epistemologisch privilege en houdt ook niet vast aan de pretentie van algemene geldigheid. De vraag is echter of het contextualisme ons niet met nog een groter probleem opzadelt, te weten: relativisme.

IV. De vruchten van ironie: inzicht verwerven in hoe we betekenis verlenen aan de wereld (gepubliceerd in: *Studies in Philosophy of Education, 25*(6), 479-496)
Het vierde hoofdstuk vloeit op twee punten voort uit het voorgaande. Allereerst vertrekt het vanuit het idee van epistemologische onzekerheid dat kenmerkend is gebleken voor de hedendaagse opvoedingsfilosofie. Het gaat uit van de uiteindelijke onbeslisbaarheid van opvoedingsfilosofische kwesties die met deze onzekerheid in verband wordt gebracht – bijv. door de antifundationalistische auteurs die zijn onderzocht in hoofdstuk twee. Ik probeer te achterhalen hoe binnen de opvoedingsfilosofie met een dergelijke onbeslisbaarheid op een verantwoorde, maar ook inhoudelijk constructieve manier kan worden omgegaan. Ik concentreer me hierbij in het bijzonder op de ironie als een opmerkelijk instrument dat filosofen op verschillende momenten hebben ingezet wanneer zij zich geconfronteerd zagen met vormen van filosofische onzekerheid of onbeslisbaarheid. De vraag die ik me hierbij heb gesteld is: wat valt te verwachten van het inzetten van ironie als filosofisch instrument in relatie tot onbeslisbaarheid en welke consequenties volgen daaruit voor de potentiële praktische relevantie van de opvoedingsfilosofie? Het hoofdstuk sluit verder aan bij het voorgaande, omdat uitgebreid wordt ingegaan op het idee van horizontaal geordende rechtvaardigingscontexten zoals dat naar voren komt in Rorty's interpretatie van ironie als filosofisch instrument. Van daaruit levert hoofdstuk vier ook inzichten op die van belang zijn voor mijn doorgaande epistemologische zoektocht.

In het algemeen wordt duidelijk dat 'ironische' filosofen, onder wie Schlegel en Kierkegaard, aangeven dat ironie eerder vragen oproept, dan dat het antwoorden oplevert. Juist in de vragen die de ironie oproept schuilt volgens deze auteurs het inzicht dat met het inzetten van ironie wordt gegenereerd. Gebaseerd op enkele recente benaderingen van ironie ontwikkel ik een eigen interpretatie, waarbij ironie een inzicht genereert in de onherleidbare wisselwerking tussen de informatieve inhoud van een bewering en de vooronderstelde communicatieve context waarin deze uitspraak wordt gedaan. Meta-ironische reflectie laat overigens zien dat deze benadering van ironie – evenals andere benaderingen – op haar beurt ook alleen als informatief begrepen kan worden tegen de achtergrond van een specifieke vooronderstelling over de manier waarop menselijke betekenisverlening tot stand komt. Voor de opvoedingsfilosofie levert dit een werkwijze op waarbij beweringen worden geanalyseerd als amendementen op communicatieve contexten zoals vooronderstelde door sprekers dan wel auteurs. Ik illustreer deze werkwijze aan de hand van discussies over de problematiek rond 'students at risk' – het probleem van groepen leerlingen die het risico lopen structureel achterop te raken op het gebied van het behalen van leerresultaten; in het Nederlands ook wel aangeduid als 'onderwijsachterstandenproblematiek'. Hoewel de 'werkelijke' context waarin de uitspraak werd gedaan nooit met zekerheid valt te reconstrueren – en dus evenmin de 'werkelijke' informatieve inhoud van de bewering –, resulteert een dergelijke analyse in een inzicht in de wisselwerking van contextuele (bij het gehoor vooronderstelde opvattingen over 'students at risk') en informatieve (suggesties over het definiëren en/of omgaan met 'students at risk') dimensies van beweringen en hun aanvaardbaarheid.

V. Epistemologische inzichten en consequenties voor de opvoedingsfilosofie II: relativisme, willekeur en dynamisch-discursieve contexten

In hoofdstuk vijf vervolg ik mijn exploratie van het contextualisme als mogelijke acceptabele epistemologische benadering voor de opvoedingsfilosofie waarin de onvermijdelijke epistemische onzekerheid van kennisclaims kan worden verdisconteerd. Mijn eerste vraag is hoe het idee van rechtvaardigingscontexten verder geconcretiseerd kan worden. Allereerst ga ik nader in op Rorty's idee van 'vocabulaires' als rechtvaardigingscontexten, zoals naar voren gekomen in de bespreking van diens interpretatie van ironie in hoofdstuk vier. Het wordt duidelijk dat Rorty's interpretatie van communicatieve rechtvaardigingscontexten aan de ene kant bruikbare inzichten oplevert, omdat hij ons toont hoe de rechtvaardiging, evaluatie en modificatie van onze manieren van spreken over de wereld begrepen kan worden, los van hoe de wereld *is*. Aan de andere kant slaagt Rorty er niet goed in duidelijk te maken hoe we enerzijds gebonden zijn aan vocabulaires, maar anderzijds vocabulaires kunnen ontstijgen, bijv. door nieuwe vocabulaires te creëren. Omdat Rorty niet goed inzichtelijk maakt hoe de transformatie, of vernieuwing van rechtvaardigingscontexten – in zijn geval begrepen als 'vocabulaires' - begrepen kan worden, blijft zijn interpretatie ook vatbaar voor het verwijt van relativisme.

Vervolgens wordt een dynamisch-discursieve contextbegrip ontleend aan hoofdstuk vier, dat me in staat stelt te verklaren hoe sprekers in communicatieve processen aan de ene kant gebonden zijn aan rechtvaardigingscontexten, maar deze contexten op het zelfde moment transformeren, waarmee ook direct aan het relativismeverwijt ontkomen kan worden. Het idee van dynamisch-discursieve contexten roept een beeld op van rechtvaardiging dat overeen komt met de wijze waarop we in alledaagse gesprekken omgaan met rechtvaardiging, maar waarvan ik betoog dat het ook gebruikt kan worden om een meer academisch-intellectueel georiënteerde epistemologie te ontwikkelen. Aldus ben ik aangekomen bij een epistemologie die ik 'discursief' noem, omdat de rechtvaardiging en ontwikkeling van kennis erbinnen wordt beschouwd als volledig afhankelijk van de voortgaande uitwisseling van claims, argumenten en bezwaren (of tegenargumenten) tussen participanten aan een communicatief proces.

Met betrekking tot de consequenties van zo'n discursieve epistemologische benadering voor de taken van de opvoedingsfilosofie concludeer ik in hoofdstuk vijf dat er in elk geval geen directe epistemologische reden meer is voor het hebben van prescriptieve pretenties, zoals uitgedragen door fallibalistische opvoedingsfilosofen. Het bepleiten van een opvoedingsfilosofie die zich beperkt tot het onder de aandacht brengen van contextuele restricties, zoals we dat tegenkomen bij de onderzochte antifundationalistische auteurs, lijkt echter ook niet vruchtbaar: in de eerste plaats vanwege het normatieve gehalte van het pleidooi, waardoor het zelf op zijn minst de schijn heeft van prescriptiviteit; ten tweede omdat zo'n pleidooi uiteindelijk zelf niet als echt contextualistisch kan worden beschouwd is. Het idee van restricties die onvermijdelijk verbonden zijn met rechtvaardiging fungeert bij deze auteurs feitelijk als een pseudo-fundament ter rechtvaardiging van een specifieke interpretatie van de opvoedingsfilosofie. Naar mijn idee benut het door deze auteurs voorgestelde model om die reden onvoldoende de mogelijkheden die worden geboden door een contextualistische benadering van de opvoedingsfilosofie.

VI. De wereld in onderhandeling: filosofische overwegingen bij het omgaan met achterstanden in de ontwikkeling van schooltaalvaardigheid (gepubliceerd in: *Educational Philosophy and Theory, 40*(5), 652-665)

Hoofdstuk zes biedt een verdere uiteenzetting van het idee van dynamisch-discursieve rechtvaardigingscontexten, waarbij speciale aandacht wordt besteed aan de activiteit van de individuele participant aan de communicatie. Deze speciale aandacht is nodig om verder te kunnen onderzoeken of de voorgestelde discursieve epistemologie niet noodzakelijk conventionalisme impliceert. De activiteit van de individuele taalgebruiker wordt onderzocht in het licht van de vraag wat het betekent voor een subject om een taal te leren gebruiken. Met het oog op de potentiële relevantie van de opvoedingsfilosofie voor bredere pedago-gische debatten, wordt bekeken of het beantwoorden van deze vraag ons kan helpen om een idee te vormen over hoe in het onderwijs het beste omgegaan kan worden met wat men in het Engels aanduidt als 'differential academic language proficiency' (vergelijkbaar met het Nederlandse debat over 'taalachterstanden in het onderwijs'), een onderwerp dat sterk samenhangt met het vraagstuk rond 'students at risk' dat in hoofdstuk zes al aan de orde is geweest.

Het onderzoek in dit hoofdstuk resulteert in het inzicht dat het leren gebruiken van een taal het beste begrepen kan worden als actieve participatie aan het proces van praktijkgerelateerde communicatie (onderhandeling) over hoe de wereld geacht wordt in elkaar te zitten, een proces dat als 'worldmaking' is aangeduid. Verschillen in '(academic) language proficiency' verschijnen dan niet als verschillen in vaardigheden in het hanteren van een linguïstisch instrumentarium, maar eerder als verschillen in expertise waar het gaat om participatie aan dit proces van 'worldmaking'. Zo bezien lijkt het aannemelijk om in het omgaan met 'differential academic language proficiency' in het onderwijs een meer substantiële rol toe te bedelen aan de gezamenlijke omgang met en communicatie over probleemsituaties – waarin dus ruimte is voor gezamenlijke betekenisconstructie –, dan tot nu toe veelal het geval is.

VII. Epistemologische inzichten en consequenties voor de opvoedingsfilosofie III: discursieve epistemologie en de groei van kennis

Voortbordurend op de inzichten uit hoofdstuk zes wordt het idee van een discursieve epistemologie verder ontwikkeld in hoofdstuk zeven. Er wordt duidelijk gemaakt hoe communicatieve processen beschouwd kunnen worden als praktijkgerelateerde onderhandeling over hoe wereld is. In die onderhandeling speelt elke participant aan de communicatie zowel de rol van initiator van de transformatie van de praktijkgerelateerde vooronderstellingen over de hoe wereld in elkaar zit, als de conservator van – een ander deel van – die verzameling van vooronderstelling. Dit beeld toont onmiddellijk dat deze benadering geen conservatief conventionalisme kan inhouden.

In een discursieve benadering van de epistemologie verschijnt de ontwikkeling van kennis als resultante van een kritisch-discursief proces waarin zelfs de meest basaal geachte vooronderstellingen op elk moment dat er aanleiding toe wordt gezien in de communicatie ter discussie gesteld kunnen worden. Aan de andere kant kunnen bepaalde vooronderstellingen het een zeer lange tijd uithouden, wanneer zij in de ogen van de participanten aan de communicatie blijven voldoen. In zo'n geval kan gesproken worden van een toestand van 'reflective equilibrium'. Echter, ook in een staat van – relatieve – stabiliteit wordt de rechtvaardigings-context voortdurend geëvalueerd, aangevuld en getransformeerd, en de mogelijkheid blijft voortdurend aanwezig dat het evenwicht onverwacht verstoord wordt, omdat sommige van de vooronderstellingen, waarvan we dachten dat ze misschien wel onbetwijfelbaar waren, onder vuur zijn komen te liggen.

Met betrekking tot de vraag naar de mogelijkheid van weten-schappelijke vooruitgang binnen een dergelijke discursieve benadering van kennis wordt verder betoogd dat van zo'n vooruitgang alleen gesproken kan worden in het licht van specifieke criteria die worden ge-hanteerd binnen een bepaalde wetenschappelijke communicatieve praktijk. Deze criteria zijn daarbij zelf ook gewoon onderdeel zijn van het voortgaande debat – het gaat immers om vooronderstelde aanvaard-baarheidsnormen, waardoor deze ook zelf op enig moment in het debat ter discussie kunnen worden gesteld. Met betrekking tot de rol die afzon-derlijke wetenschappers spelen binnen de wetenschappelijke com-municatieve praktijk wordt in dit verband geconcludeerd dat wetenschap-pelijke expertise niet alleen verwijst naar het vermogen om goed onder-zoek te doen in lijn met de algemeen geldende normen van 'goede wetenschap', maar ook om succesvol bij te dragen aan de voortdurende kritische discussie over uitgangspunten die met zo'n idee van 'goede wetenschap' zijn verondersteld en die doorgaans misschien worden beschouwd als deel uitmakend van de kern van ons actuele begrip van wat 'wetenschap' inhoudt.

VIII. Het primaat van engagement binnen de opvoedingsfilosofie

Hoofdstuk twee liet eerder zien dat waar het rechtvaardiging van opvoedingsfilosofische uitspraken betreft, hedendaagse antifundationa-listische opvoedingsfilosofen uitgaan van een 'primaat van engagement'. Bij deze auteurs houdt dit in dat de wijze waarop zij inrichting geven aan

177

de opvoedingsfilosofie onmiddellijk samenhangt met een praktisch engagement met het bestrijden van uitsluiting. In hoofdstuk acht onderwerp ik dit idee aan een nader onderzoek. Ik betoog in de eerste plaats dat de afwijzing van het fundationalistische rechtvaardigingsmodel inderdaad een 'primaat van engagement' impliceert met betrekking tot rechtvaardiging in het algemeen. Als het onmogelijk is om ook maar een relatief zekere – algemeen geldende – epistemologische basis voor rechtvaardiging te formuleren, dan is de opvoedingsfilosoof – evenals enige andere wetenschapper, of welke spreker dan ook – uiteindelijk altijd teruggeworpen op bepaalde voorkeuren, overtuigingen en affiniteiten. Ik onderzoek verder wat zo'n 'primaat van engagement' precies in kan houden en wat voor consequenties voor de taken en mogelijkheden van de opvoedingsfilosofie ermee verbonden zijn, waarbij ik in het bijzonder ook inga op haar praktische relevantie.

Ik kom tot de conclusie dat wijze waarop het 'primaat van engagement' wordt begrepen door de onderzochte antifundationalistische auteurs problematisch is, omdat ze de oorsprong van het engagement uiteindelijk toch buiten de communicatie plaatsen; dan wel in een pre-talig affect, dan wel in gedeelde conventies. Gebruik makend van de eerder verworven inzichten over rechtvaardiging binnen dynamische communicatieve contexten betoog ik dat het 'primaat van engagement' het beste begrepen kan worden als het idee dat de ontwikkeling en rechtvaardiging van kennis uiteindelijk is te herleiden tot een proces waarin een geëngageerde proponent een claim formuleert en bereid is te verdedigen ten overstaan van een specifiek gehoor, zoals door deze proponent waargenomen of voorgesteld. En het is in de bereidheid van de proponent om deze specifieke claim uit te dragen en te verdedigen waarin het engagement schuilgaat.

Deze interpretatie van het 'primaat van engagement' vraagt ook om een andere benadering van de vraag hoe uitsluiting bestreden kan worden, die blijkbaar een belangrijke rol speelt in hedendaagse opvoedingsfilosofische debatten. Bezien vanuit de hier ontwikkelde benadering kan het bestrijden van uitsluiting het beste begrepen worden in termen van het ter discussie stellen van praktijkgerelateerde normen die reguleren wie al dan niet wordt geaccepteerd als actieve participant aan de communicatie.

IX. Samenvattende en concluderende opmerkingen: engagement en academische striktheid in de opvoedingsfilosofie

In dit laatste hoofdstuk van het proefschrift geef ik eerst een samenvatting van de belangrijkste inzichten uit de voorgaande zeven hoofdstukken (hoofdstuk 2 t/m 8), om vervolgens in te gaan op de drie onderzoeksvragen die in hoofdstuk één zijn geformuleerd en die richtinggevend zijn geweest voor de doorgaande argumentatieve lijn binnen dit proefschrift.

Met betrekking tot de vraag hoe we het idee van de zogenaamde 'onzekerheid van kennis' kunnen verdisconteren in een filosofisch acceptabele epistemologische benadering concludeer ik dat deze te vinden is in wat ik een 'discursieve' benadering van de epistemologie heb genoemd. Binnen zo'n benadering wordt het in epistemische zin gerechtigd zijn een bepaalde claim te doen – oftewel, het aanspraak kunnen maken op 'kennis' – begrepen in termen van de acceptatie van een communicatieve bijdrage van een spreker door een bepaald publiek. Van zo'n situatie kan alleen sprake kan zijn als het publiek ook denkt dat de spreker ook voldoende goede redenen heeft voor het doen van deze claim. Zo niet, dan zal de spreker gevraagd worden zijn claim verder te onderbouwen, of zal de claim zelfs direct verworpen worden. Kennis verschijnt zo als de praktijkgebonden vooronderstelde 'common ground' – de communicatieve context – waarop in een communicatief proces een beroep moet worden gedaan om communicatie überhaupt mogelijk te maken, maar die ook met het doen van elke communicatieve bijdrage, althans delen ervan, ter discussie wordt gesteld.

Wetenschap wordt zo bezien niet geacht een exclusieve, of in enigerlei wijze geprivilegieerde toegang te hebben tot kennis over de wereld, hetgeen overigens geenszins impliceert dat aan wetenschap geen speciale status toegekend kan worden als kennisautoriteit. Of dit zo is, hangt af van de waarde die binnen een gemeenschap aan wetenschap wordt toegekend. De opvoedingsfilosofie kan tegen deze achtergrond beschouwd worden als onderscheiden wetenschappelijke communicatieve praktijk met eigen kennisaanspraken en eigen kenniscriteria, waarbij die criteria zelf ook direct onderwerp zijn van het opvoedingsfilosofische debat.

Waar het de vraag naar de taakstelling van de opvoedingsfilosofie betreft, kom ik tot de conclusie dat er geen directe epistemologische aanleiding is voor de opvoedingsfilosofie om zich louter te beperken tot kritiek in de vorm van het onder de aandacht brengen van contextuele restricties aan rechtvaardiging– zoals werd gesuggereerd door antifundationalistische opvoedings-filosofen. Mijn ironische analyse van wetenschappelijke publicaties over het omgaan met 'students at risk' in hoofdstuk vier heeft echter laten zien dat het reconstrueren van contextuele restricties wel degelijk waardevol kan zijn, enerzijds omdat het een bijdrage levert aan het kritisch-reflectieve vermogen van de opvoedings-filosofie en anderzijds omdat het een inzicht genereert in de wijze waarop we betekenis verlenen aan de wereld – in dit geval de wereld van opvoeding en onderwijs. Mijn interpretatie van het leren gebruiken van een taal

179

door een subject als een actieve participatie aan een voortgaande praktijk-gebonden onderhandeling over de wereld toont verder dat ook een theoretisch-inhoudelijke bijdrage van de opvoedingsfilosofie verwacht kan worden, ook al wordt er geen beroep gedaan op het toekennen van epistemologisch privilege en wordt de pretentie van algemene geldigheid losgelaten. Wel ligt het in dat licht voor de hand om terughoudendheid te betrachten waar het prescriptieve pretenties betreft.

Terugkijkend op het proefschrift als een voorbeeld van een (collectie van) bijdrage(n) aan de opvoedingsfilosofische communicatieve praktijk, wordt duidelijk gemaakt dat een scala aan taken voorstelbaar is voor een opvoedingsfilosofie die zich rekenschap geeft van het feit dat kennis altijd tot een bepaalde hoogte onzeker is. Hoe de opvoedings-filosofie zich ontwikkelt, zo wordt tot slot betoogd, zal uiteindelijk te herleiden zijn tot het engagement van de participanten aan het opvoe-dingsfilosofisch debat. Daarbij wordt onmiddellijk geruststellend gecon-cludeerd dat een dergelijk 'primaat van engagement' in de opvoedings-filosofie de wetenschappelijke strengheid van de discipline geenszins aan-tast, sterker nog: ze is vormt er een voorwaarde voor.

Dit brengt ons bij de vraag naar de mogelijke praktische relevantie die te verwachten valt van de opvoedingsfilosofie. Hoewel een mogelijke impact van opvoedingsfilosofische inzichten op meer praktische discus-sies over opvoeding en onderwijs zeker tot de mogelijkheden behoort, wordt wel duidelijk gemaakt dat het bezien vanuit mijn discursieve benadering volledig buiten de invloedssfeer van de opvoedingsfilosofie zelf ligt óf van zo'n impact sprake zal zijn en wat die impact dan zou moeten zijn. Tegen die achtergrond wordt betoogd dat een opvoedings-filosofie die haar eigen bestaan (mede) legitimeert in termen van haar praktische relevantie problematisch is.

DANKWOORD (ACKNOWLEDGEMENTS IN DUTCH)

Het kan niet genoeg benadrukt worden dat er veel mensen zijn die – al dan niet bewust – een rol hebben gespeeld bij de totstandkoming van dit proefschrift. Hierbij wil ik iedereen die een bijdrage heeft geleverd aan de afronding van het promotietraject, of die de tijd die ervoor nodig was draaglijker heeft gemaakt van harte bedanken.

Als eerste wil ik me richten tot Frieda Heyting die als promotor mij de gelegenheid heeft geboden dit promotietraject in te gaan en die mij tot het allerlaatste moment zeer intensief heeft begeleid. Frieda, wat ik van jou heb geleerd van het moment dat ik student was tot aan de laatste contacten rond de afronding van het proefschrift reikt veel verder dan ons vakgebied. Het arbeidsethos en morele verantwoordelijkheidsbesef dat jij aan de dag hebt gelegd in de uitoefening van je vak heeft mij altijd enorm geïnspireerd. Dat je zelfs tot ver in je emeritaat bereid bent geweest om zoveel tijd en energie te steken in de begeleiding van mijn onderzoek is heel bijzonder en bewonderenswaardig. Voor al deze zaken, het vertrouwen dat je in mij hebt gesteld en de persoonlijke steun die ik heb ervaren bij de afronding van dit zware proces kan ik je niet genoeg bedanken. Jij bent het levende bewijs dat pragmatisme geen onverschilligheid impliceert. Dat je niet als promotor vernoemd mag worden omdat je nu te lang met emeritaat bent spijt me enorm.

Mijn dank gaat ook uit naar Michael Merry, mijn (nieuwe) promotor. Om pas in een laat stadium als promotor bij een promotieonderzoek betrokken te raken is gecompliceerd, zeker als de promovendus van mening is dat het werk al zo goed als gedaan is. Michael, bedankt dat je bereid bent geweest om deze verantwoordelijkheid toch op je te nemen. Jouw kritische commentaar en onze discussies hebben mijn ogen geopend voor mogelijke bezwaren van inhoudelijke 'tegenstanders', hetgeen uiteindelijk ook heeft geresulteerd in een duidelijke verbetering van het proefschrift.

De leden van de promotiecommissie dank ik heel hartelijk voor de bereidheid om energie te steken in het lezen en beoordelen van het proefschrift en tijd vrij te maken om de promotieplechtigheid te kunnen bijwonen.

Verder bedank ik de medeopvoedingsfilosofen waarmee ik gedurende het onderzoeksproces in gesprek ben geweest over de opvoedingsfilosofie in het algemeen en mijn onderzoek in het bijzonder. Binnen de UvA ging het dan om Gert-Jan Vreeke, Jantine Hemrica en Katka Lepková. Gert-Jan, jij bent erg belangrijk geweest. In onze gesprekken kon ik mijn gedachten toetsen en aanscherpen. Je fungeerde daarnaast vaak als welkome participant-mediator in de soms heftige discussies met

181

Frieda over het onderzoek. Maar bovenal heb je ervoor gezorgd dat ik het vertrouwen niet kwijtraakte als er stevige kritiek was op stukken die ik geschreven had. Jantine en Katka, zonder medepromovendi die dezelfde moeilijkheden als ik hebben ervaren had ik niet alleen lief en leed minder goed kunnen delen, maar had ik ook zeker minder gelachen. Buiten de UvA zijn de leden van het Vlaams-Nederlandse Kohnstammnetwerk voor wijsgerige en historische pedagogiek belangrijk geweest. Ik ben verschillende leden erkentelijk voor hun kritisch commentaren op stukken die ik had geschreven. Daarnaast denk ik met heel veel plezier terug aan de minder formele momenten op en rond bijeenkomsten of congressen in Utrecht, Leuven, Madrid en Oxford. Onder het genot van wat drankjes met vakgenoten (vaak tot diep in de nacht) filosoferen, associëren, discussiëren en relativeren werkt niet alleen ontspannend, maar ook louterend.

Een aantal collega's die er mede voor hebben gezorgd dat ik het altijd goed naar mijn zin gehad op de UvA wil ik met name noemen. In de eerste plaats de historici binnen de vakgroep grondslagen en geschiedenis: Bernard Kruithof, Hugo Röling en in het bijzonder Ernst Mulder waarmee ik altijd heel prettig intensief heb kunnen samenwerken. Daarnaast wil ik (destijds) medepromovendi Sophie, Reinoud, Jessica, Roos, Inge, Hester en Femke speciaal danken voor de plezierige momenten die we met elkaar hebben mogen beleven.

Hogeschool Inholland dank ik voor de ondersteuning in de laatste fase van het onderzoek. Mijn dank gaat met name uit naar de mensen die deze ondersteuning hebben mogelijk gemaakt: Brecht van Schendel, Ingrid Feenstra, Lisette van der Poel, Janny van Tuyl en Marij Urlings.

In het laatste gedeelte van dit dankwoord besteed ik aandacht aan de mensen in mijn persoonlijke leven die direct of indirect een bijdrage hebben geleverd, maar die ook in meer of mindere mate last hebben gehad van de worsteling die het schrijven van het proefschrift ook vaak was.

In de eerste plaats bedank ik al mijn vrienden in Amsterdam, Tilburg en Limburg, gewoonweg voor het feit dat ze vrienden zijn. Het ontroert mij regelmatig dat het blijkbaar mogelijk is om in je leven zulke fijne mensen om je heen te verzamelen. Met name noem ik mijn paranimfen Albert (want Appie mag niet meer), Rogier en Bas. Jullie maken al zo lang zo'n belangrijk deel uit van mijn leven; het doet me erg goed dat jullie deze rol willen vervullen. Verder moeten Mirjam en Stella genoemd worden, omdat ik mijn belofte dat zij paranimf mochten zijn niet ben nagekomen. Wat fijn dat wij vrijwel nooit over het proefschrift hoefden te praten. Jullie zijn echt fijne vriendinnetjes. De Takkeflikkers noem tot slot ook expliciet, omdat ik zeker gezeur krijg als ik dat niet doe. Mannen, hierbij wederom een vermelding in een proefschrift!

Als pedagoog kun je de invloed van het 'nest' waar je ooit

uitgevlogen bent natuurlijk niet over het hoofd zien. Pap en mam, jullie hebben mij altijd mijn eigen keuzes laten maken en steunen mij nog altijd bij de dingen die ik doe. Dat biedt een basis van vertrouwen en de mogelijkheid te ontdekken wat je kan en wat je wil. Dat ik de wil en durf heb gehad om aan dit project te beginnen en dat ik in staat ben geweest om het af te ronden heb ik voor een groot deel aan jullie te danken. Edwin, jij was de grote broer op de universiteit, dus uiteindelijk moest ik dat ook. De inhoudelijke discussies die we van jongs af aan hebben gevoerd hebben ontegenzeggelijk mijn denken gescherpt en mijn argumentatief vermogen mede gevormd.

Eliska, van iedereen heb jij waarschijnlijk het meeste last gehad van alle negatieve kanten die ook aan het schrijven van dit proefschrift zaten. Vaak geen leuke avonden, weekenden, of vakanties, omdat ik dacht dat ik dan flink kon opschieten. Maar meer nog waarschijnlijk de passiviteit, negativiteit en afstandelijkheid, voortkomend uit schuldgevoel als ik toch weer minder had gedaan dan ik me had voorgenomen. Het is moeilijk te zeggen hoe dat alles onze relatie heeft beïnvloed. Wel kan ik zeggen dat ik heel blij ben dat wij elkaar ooit hebben leren kennen en jij weet dat dat niet alleen zo is omdat we samen zo'n geweldig dochter hebben. Je bent een geweldig mens en een hele toffe moeder.

Annelies, de eerlijkheid gebied te zeggen dat dit proefschrift er waarschijnlijk niet had gelegen als wij elkaar niet waren tegengekomen in 'de Tramhalte' op die carnavalszondag. Het is verbazingwekkend hoeveel rust en ik ervaar sinds wij samen zijn. Ik houd heel veel van je en heb veel zin om met jou allemaal nieuwe stappen te gaan zetten.

Lize, dit proefschrift is opgedragen aan jou. Jij hebt er vaak van gebaald als ik weer aan mijn 'boekje' moest werken. Dat jij er bent heeft mij toch ook juist geïnspireerd om door te blijven gaan met schrijven en er in elk geval alles aan te doen om tot een afronding te komen. Ik verwacht niet dat je het 'boekje' ooit zult lezen, maar misschien doet alleen al het feit dat het geschreven is 'voor Lize' je in de toekomst af en toe glimlachen. Je bent een geweldige dochter!

Printed and bound by CPI Group (UK) Ltd, Croydon, CR0 4YY

27/10/2024

14580699-0003